SOCIETY FOR EXPERIMENTAL BIOLOGY
SEMINAR SERIES · 8

STOMATAL PHYSIOLOGY

STOMATAL PHYSIOLOGY

Edited by

P. G. JARVIS

Professor of Forestry and Natural Resources, University of Edinburgh

AND

T. A. MANSFIELD

Professor of Biological Sciences, University of Lancaster

CAMBRIDGE UNIVERSITY PRESS

Cambridge

London   New York   New Rochelle

Melbourne   Sydney

CAMBRIDGE UNIVERSITY PRESS
Cambridge, New York, Melbourne, Madrid, Cape Town, Singapore, São Paulo

Cambridge University Press
The Edinburgh Building, Cambridge CB2 2RU, UK

Published in the United States of America by Cambridge University Press, New York

www.cambridge.org
Information on this title: www.cambridge.org/9780521236836

© Cambridge University Press 1981

First published 1981

A catalogue record for this publication is available from the British Library

ISBN-13 978-0-521-28151-5 paperback
ISBN-10 0-521-28151-2 paperback

Transferred to digital printing 2006

CONTENTS

CONTRIBUTORS

Allaway, W. G.
School of Biological Sciences, The University of Sydney, NSW 2006, Australia.

Ayres, P. G.
Department of Biological Sciences, University of Lancaster, Lancaster LA1 4YQ, UK.

Black, C. R.
Department of Physiology and Environmental Studies, University of Nottingham School of Agriculture, Sutton Bonington, Loughborough, Leicestershire LE12 5RD, UK.

Black, V. J.
Department of Physiology and Environmental Studies, University of Nottingham School of Agriculture, Sutton Bonington, Loughborough, Leicestershire LE12 5RD, UK.

Davies, W. J.
Department of Biological Sciences, University of Lancaster, Lancaster, LA1 4YQ, UK.

Jarvis, P. G.
Department of Forestry and Natural Resources, University of Edinburgh, The King's Buildings, Mayfield Road, Edinburgh EH9 3JU, UK.

Jarvis, R. G.
Department of Biological Sciences, University of Lancaster, Lancaster, LA1 4YQ, UK.

Lösch, R.
Lehrstuhl für Botanik II, Botanisches Institut der Universität Wurzburg, 8700 Wurzburg, BRD.

MacRobbie, E. A. C.
Botany School, University of Cambridge, Downing Street, Cambridge CB2 3EA, UK.

Mansfield, T. A.
Department of Biological Sciences, University of Lancaster, Lancaster LA1 4YQ, UK.

Meidner, H.
Department of Biology, The University, Stirling, FK9 4LA, UK.

Morison, J. I. L.
Department of Forestry and Natural Resources, University of Edinburgh, The King's Buildings, Mayfield Road, Edinburgh EH9 3JU, UK.

Osonubí, O.
Department of Biological Sciences, University of Lancaster, Lancaster LA1 4YQ, UK.

Palevitz, B. A.
Department of Botany, University of Georgia, Athens, Georgia 30602, USA.

Sharp, R. E.
Department of Biological Sciences, University of Lancaster, Lancaster LA1 4YQ, UK.

Squire, G. R.
Department of Physiology and Environmental Studies, University of Nottingham School of Agriculture, Sutton Bonington, Loughborough, Leicestershire LE12 5RD, UK.

Tenhunen, J. D.
Lehrstuhl für Botanik II, Botanisches Institut der Universität Wurzburg, 8700 Wurzburg, BRD.

Travis, A. J.
Department of Biological Sciences, University of Lancaster, Lancaster LA1 4YQ, UK.

Unsworth, M. H.
Department of Physiology and Environmental Studies, University of Nottingham School of Agriculture, Sutton Bonington, Loughborough LE12 5RD, UK.

Willmer, C. M.
Department of Biology, University of Stirling, Stirling FK9 4LA, UK.

Wilson, J. A.
Department of Biological Sciences, University of Lancaster, Lancaster LA1 4YQ, UK.

Zeiger, E.
Department of Biological Sciences, Stanford University, Stanford, California 94305, USA.

PREFACE

The contributors to this volume were the invited speakers in a seminar organized by the Society for Experimental Biology at the University of Lancaster on 19–20 December 1979. The object of the seminar was to present the many facets of a subject that has assumed increasing importance over the past two decades. Stomatal physiology is now covered in some depth in many undergraduate courses, and is encountered by many graduate students and researchers whose main interests are in other areas of plant biology. It is hoped that this collection of papers will be helpful to many people in these categories. Our contributors have not only presented results of their own recent research, but have introduced their topics for the non-specialist.

B. A. PALEVITZ

The structure and development of stomatal cells

Introduction

Botanists have long been intrigued by the relationship between structure and function in plant cells, including those that comprise the stomatal apparatus. Because the architecture and content of guard cells is so obviously important in their behaviour, it is pertinent to inquire into their structure and the mechanisms that govern their differentiation. Furthermore, a thorough understanding of the structure and development of guard cells can only be obtained with comprehensive information at the cell biological level. Such information can assist us in interpreting how developing cells respond to external signals emanating from the environment and from other cells, and what internal processes control differentiation.

Guard cells are useful as model systems for the study of basic problems in cell biology and development. The processes of growth and cell shaping, and the 'cooperative' interaction between adjacent cells during development and function, have long intrigued biologists. These can be successfully studied using stomatal cells. Guard cells in particular offer favourable material to attack problems such as the control of osmotic and ionic fluxes at membranes, the control of cell surface activity during the deposition of extracellular macromolecules (cell wall), and the role of motile or cystoskeletal elements (microtubules and microfilaments) in the regulation of plasma membrane activity and wall formation.

The subject of stomatal structure has been reviewed recently (Allaway & Milthorpe, 1976). In this article, I shall review various aspects of stomatal anatomy, with an emphasis on wall formation in developing guard cells. The reader should gain an appreciation for the timing and relationships of various formative processes to each other during maturation. Hopefully, the reader will also sense the value of stomata in obtaining information of more general significance at the cellular level.

Cell division

The formation of stomatal cells involves the precise placement of division planes in parent cells. The mechanisms that determine the orientation of the new cell plate in plants have always been of intense interest, and much attention has been directed at the stomatal apparatus as a model system. The formation of guard mother cells and subsidiary cells usually involves asymmetric division in which the nucleus migrates to a specific site in the parent cell before dividing (reviewed by Hepler & Palevitz, 1974). In addition, the orientation of the new cell plate is often radically different from that resulting from the division of other epidermal cells. Little is known about the mechanisms underlying these phenomena. Considerable circumstantial evidence indicates that cells such as guard mother cells somehow control mitotic activity and the placement of division planes in neighbouring cells (e.g. subsidiary cells) by some field effect involving chemical gradients. Undoubtedly motility-producing macromolecules (microtubules, actin) are involved in the requisite movements, but which ones operate in each case is unclear. It is known that certain inhibitors such as colchicine and cytochalasin B may interfere with nuclear migrations and polarity phenomena (Hepler & Palevitz, 1974). In the division of guard mother cells of *Allium cepa* L., the spindle apparatus is oriented obliquely through anaphase, leading to the separation of daughter chromosomes into opposite corners of the cell (Palevitz & Hepler, 1974; Palevitz, 1980). A specific, directed movement of the daughter nuclei and forming cell plate then ensues, leading to the proper longitudinal placement of the plate. It appears that both microtubules and microfilaments are involved, because the process is sensitive to agents which affect both structures (colchicine, cytochalasin B, phalloidin). This same process also occurs in the guard mother cells of other *Allium* species as well as in those of the Gramineae. A preprophase band of microtubules is known to mark the future division plane in these cells, but a role in actually orienting the cell plate does not seem likely because the band disappears long before telophase. Instead, the band may reflect a circumferential, morphogenetically-determined region of the plasma membrane that somehow interacts with the nucleus, spindle apparatus or forming cell plate during division (Hepler & Palevitz, 1974; Palevitz & Hepler, 1974). French & Paolillo (1975) reported that the plane of division in guard mother cells of moss sporophytes is influenced by physical contact with the calyptra. The significance of this observation in relation to the above movements is unclear. It is hoped that a solution to these long-perplexing processes will soon be forthcoming.

Cell shape and wall deposition in guard cells

Perhaps the two most distinctive features of guard cells are their characteristic shapes and non-uniform wall thickenings. In plants, cell shape is governed by the surrounding cell wall. The mechanical properties of the wall in turn are in large part determined by the orientation of newly deposited cellulose microfibrils. In addition, a characteristic of many highly differentiated plant cells such as guard cells and tracheal elements is the deposition of thick, often distinctly patterned secondary walls in which the cellulose is also highly oriented. Evidence indicates that the pattern of wall deposition and the orientation of wall microfibrils in plant cells may be determined by the pattern and co-orientation of cortical microtubules adjacent to the plasma membrane (Hepler & Palevitz, 1974). Microtubules may exert this control by interacting with specific components in the plasma membrane, either indirectly through the active generation of shear in the plane of the fluid membrane or directly by linkages between microtubules and membrane-bound components such as cellulose-synthesizing complexes.

The mechanism governing the alignment of wall cellulose, however, has remained controversial, and co-orientation of microtubules and cellulose has not always been seen. One of the main reasons for lack of resolution of the problem in plants has been the choice of cells studied. What is clearly needed is a cell system which exhibits the combined characteristics of growth accompanied by shape change; the deposition of localized wall thickenings containing cellulose microfibrils, the orientation of which is precise and known; and ready accessibility enabling easy observation with light and electron microscopy. Guard cells fulfil all these criteria. In addition, a comprehensive literature exists on the physiology of these cells during gas exchange.

The patterns of wall thickening in guard cells vary amongst species. Typically the upper and lower paradermal walls in the vicinity of the pore are heavily thickened, but the amount of deposition on each can vary. The ventral wall bordering the pore is also thickened, though less so than the paradermal walls. The ventral and anticlinal end walls in some guard cells may also be thickened near their juncture. The common wall of the guard and subsidiary cells is relatively unthickened, as are other walls of subsidiary cells. In the elliptical or kidney-shaped guard cells of plants such as *Allium* and *Pisum*, the paradermal walls are thickened in a fan-like pattern that radiates away from the pore (Singh & Srivastava, 1973; Palevitz & Hepler, 1976). The ventral wall bordering the pore is also thickened. Wall microfibrils in these thickenings are oriented in a similar, radial manner; that is, they too radiate away from the pore site in a fan-like pattern (Fig. 1*a, b*).

4   B. A. PALEVITZ

Immediately following division of the precursor guard mother cell, the immature guard cells begin a marked change in shape. The first sign of differentiation in these cells is the assembly of microtubules at a site close to the plasma membrane and adjacent to the future pore in the common wall between sister guard cells (Palevitz & Hepler, 1976). Soon after the microtubules appear, the common wall thickens at this site. The new cellulose

Fig. 1. (a) Tangential section through a guard cell of *Allium cepa*. Note the thickened walls, the cuticle, and the radially-arranged microfibrils and microtubules (P, pore). Micrograph by M. Doohan & B. Palevitz. ×14300. (b). Polarization micrograph of similar *Allium* guard cells showing birefringence of aligned microfibrils. ×1000.

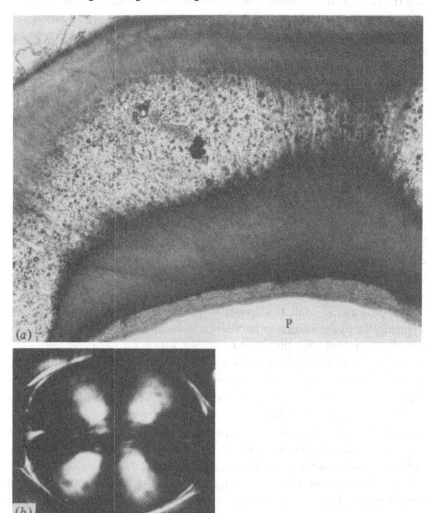

microfibrils are deposited in the radial, fan-like array described above. This alignment exactly parallels that of the population of cortical microtubules which appeared prior to wall thickening (Fig. 1*a*). Moreover, this orientation remains constant during differentiation, a point to be remembered later when we discuss the guard cells of grasses. Thus, it is proposed that microtubules determine both the localization of thickening and the orientation of cellulose. Soon after the initiation of the new wall layer, the cell begins to assume its characteristic kidney shape. The change in shape is produced by the continued growth of the cells constrained by the new, radial array of wall microfibrils. Colchicine, which leads to the breakdown of cortical microtubules, disrupts cell shape, the localization of wall thickening, and the orientation of the wall microfibrils. Thus, one of the early, determinative events in the differentiation of the guard cells of *Allium* is the precise, oriented assembly of microtubules at specific sites in the cell.

The guard cells of grasses such as timothy-grass (*Phleum pratense* L.) and maize are quite different from those of *Allium* (Kaufman, Petering, Yocum & Baic, 1970; Srivastava & Singh, 1972; Ziegler, Schmueli & Lange, 1974; Palevitz & Alones, 1977). At maturity, these cells are highly elongate and bone- or dumb-bell-shaped (Fig. 2*a*, *b*). The wall around the constricted midzone of each cell is heavily thickened (Fig. 2*a*, *c*). Early work, especially that by Ziegenspeck (1944), demonstrated that wall microfibrils in this region are axially oriented compared to the radial arrangement found in *Allium*. Because the shape and wall architecture in the guard cells of grasses is so different from that of elliptical cells, it would be instructive to compare differentiation in these species. Furthermore, in the grasses, the differentiation of guard cells is accompanied by the formation of adjacent subsidiary cells, and the two cell types function together in mature stomatal activity. It is of interest to know what effect if any such subsidiary cells have on neighbouring, growing guard cells and to what extent these cells interact during development.

Young guard cells of grasses, soon after division of the guard mother cells, have a characteristic shape – that is, their lateral walls curve inwards (Fig. 3*a*). This shape is even evident in the guard mother cells prior to division. Gradually, a pad-like thickening develops in the common wall between the guard cells in a manner similar to that of *Allium*. Electron microscopy at this early stage reveals ultrastructural detail also similar to that of *Allium*. The assembly of microtubules in a zone adjacent to the plasma membrane precedes the formation of this thickening (Fig. 5*a*). Both microtubules and wall cellulose radiate away from the site in a fan-like array as in *Allium*.

Guard cells continue to increase in size but retain their shape up to the

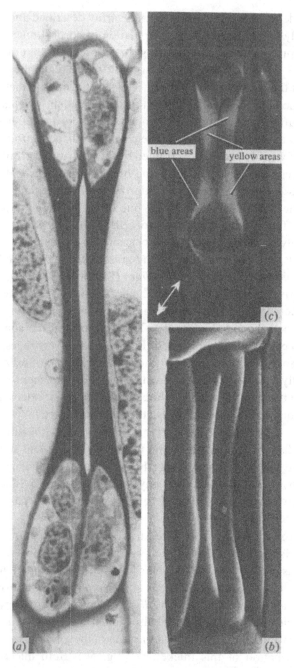

Fig. 2. (a). High voltage electron micrograph of a mature stomatal complex of timothy-grass. Note the slit-like pore and heavily thickened wall in the mid-zone of each bone-shaped guard cell. ×3500. (b). A scanning electron micrograph of a similar timothy-grass stomatal

time the pore is first formed. Then, a radical change in cell shape occurs at the time the pore first becomes evident with the light microscope. The lateral wall of the guard cells bulges out drastically as the cells assume an elliptical shape (Figs. 3b, 8). At the same time, the subsidiary cells flatten. Scanning electron microscopy as well as cross-sections of embedded cells show that the upper and lower paradermal walls bow out as well at this stage (Fig. 4). As the guard cells begin to bulge, there is a marked increase in the size of their vacuoles, indicating that active osmotic processes are responsible for the shape change (Figs. 3b, 8b). Electron microscopy shows that microtubules are still oriented radially, as in *Allium* (Fig. 5b). Indeed, the cells are now strikingly similar in overall appearance to the guard cells of *Allium*. Yet an important difference is also evident. At the time the pore opens, the new

Fig. 3. Representative sequence of differentiation in grass stomatal cells. (a) Very young cells of maize. ×1500. (b) Transient guard cell swelling stage in maize. ×1700. (c) Reconstruction of guard cells in timothy. ×1500. (d) Mature complex of timothy. ×1300.

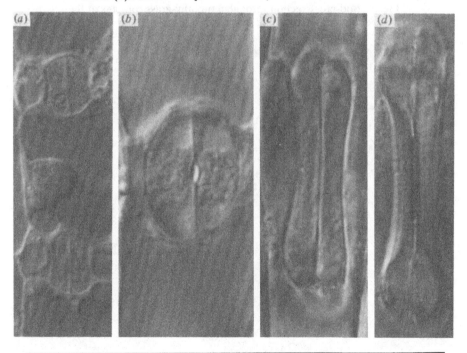

Fig. 2 continued
complex. Micrograph by M. Ledbetter & B. Palevitz. ×1200.
(c) Polarization micrograph of timothy-grass using a Red I Plate. By rotating the specimen relative to the plate major axis (bar), the resulting colours reveal wall microfibril orientation. ×1100.

wall is not nearly as thick as in *Allium* at the equivalent stage. Wall birefringence is also still hardly detectable, in contrast to *Allium*.

As stated above, grass guard cells are highly elongated and bone-shaped at maturity. But at this intermediate stage of differentiation, they assume a kidney shape. In fact, this kidney-like swelling is just a transient shape-change. The cells soon begin to reconstrict and elongate as the pore becomes long and slit-like (Fig. 3c).

What is happening to the cell wall at this time? We have already noted that the wall is not very thick in guard cells at the time of the transient shape-change. However, polarization optics shows a rapid, accelerated deposition of new wall material in the narrow mid-zone of each guard cell soon after reconstriction (Fig. 3d). This new wall layer flares out at the bulbous ends of the cell (Figs. 2a, 3d). The wall becomes very thick and birefringence retardation values eventually exceed those of *Allium*. Wall thickening is

Fig. 4. Mid-cross-section through a timothy-grass complex at the transient guard cell swelling stage. ×5300.

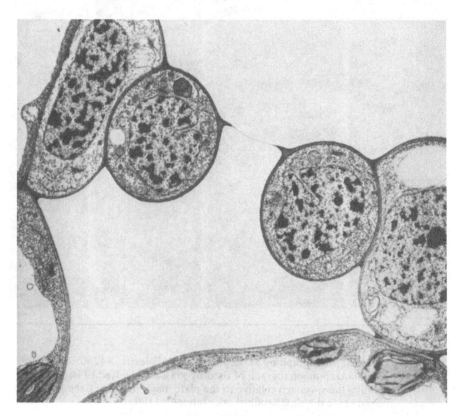

accompanied by increased activity of the Golgi apparatus and the appearance of large numbers of vesicles in the cortical cytoplasm.

What about microtubule and microfibril orientation at this stage? As I have already mentioned, Ziegenspeck (1944) claimed that the cellulose in mature grass guard cells is axial in orientation; that is, it is aligned along the long axis of the cell and not radially as in *Allium* and other species with elliptical guard cells. Electron microscope and polarization analysis shows that indeed the cellulose deposited *late in differentiation* is *axial*. More precisely, the microfibrils are net axial – i.e., they are arrayed in a steep, criss-crossed pattern (Fig. 6*a*) which flares out at the bulbous ends of the cell. Most importantly, the cortical microtubules in such cells are also oriented in a steep, axial pattern (Fig. 6*a*). Cross-bridges can also be seen linking microtubules to the plasma membrane (Fig. 6*b*).

Fig. 5. (*a*). Mid-longitudinal section soon after cell division in timothy-grass. Note microtubules (arrows) in each guard cell near the plasma membrane bordering the common wall. ×39000.
(*b*). Mid-transient guard cell in tangential section. Note the mostly radial microtubules and wall microfibrils. A cuticle is not yet present. ×32000.

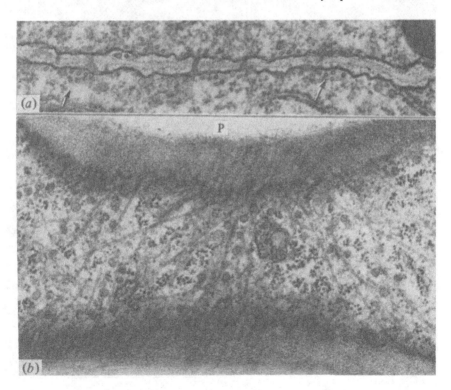

Our results above indicate that in grasses there is a shift or reorientation in microtubules, followed by cellulose, from radial to near axial, midway through differentiation when wall deposition rapidly accelerates. Indeed microtubules do seem to assume a more axial orientation just following the transient (i.e., in reconstricting cells). At first a few axial microtubules appear interspersed with the pre-existing radial array. Gradually more tubules appear in the new orientation as radial tubules decrease in number until the fully realigned arrangement is produced (Fig. 7). Thus, the shift in microtubule orientation precedes and/or accompanies the shift in cellulose orientation in the new wall layers deposited after the transient.

That the shift in microtubule orientation is morphogenetically significant

Fig. 6. (*a*). Longitudinal section through the mid-zone of a nearly mature guard cell of timothy-grass. Note the sublayers in the wall and the criss-crossed, near-axial wall microfibrils and cortical microtubules. ×50000. (*b*). An axial microtubule close to the plasma membrane in a guard cell of timothy-grass. Possible bridging elements are evident (arrows). ×150000.

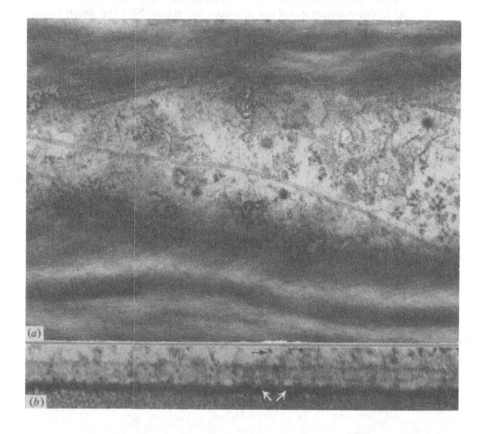

is indicated by the fact that it also occurs in subsidiary cells. Microtubules shift from a circumferential to an axial orientation concurrent with the shift in the guard cells. Thus the signals responsible for this shift are exerted across the four cells of the stomatal complex. However, no change in microtubules is evident in other epidermal cells adjacent to the stomatal complex.

The transient stage does not seem to be an artifact. It occurs in all the species examined including timothy-grass, oats, wheat, barley, and maize. Moreover, the transient is represented in the early work of Ziegenspeck (1944), Flint & Moreland (1946) and others. Examination of leaf segments stained histochemically for the presence of potassium and chloride shows that little precipitate is evident in stomatal cells prior to the transient. As this stage is entered, however, stain deposit for potassium and chloride ion becomes massive, especially in guard cells (Fig. 8*a*). Staining decreases again during reconstruction, except for precipitate which remains in subsidiary cells. The same pattern of sudden heavy staining in the transient stage was seen with neutral red (Fig. 8*b*). Guard cells in the transient develop intense, orange-pink colour in their vacuoles. Thus, active osmotic processes occur at the time of the transient and the resulting swelling of the guard cells may be necessary to pull open the pore initially.

The successful culture of grass leaf-segments has also provided insights into the transient stage and its possible relation to wall deposition. In primary leaves of timothy-grass excised under sterile conditions, slit lengthwise and placed in a drop of culture medium in Sykes–Moore

Fig. 7. Diagrammatic representation of changes in cell shape, wall deposition and microtubule–microfibril orientation in grass guard cells.

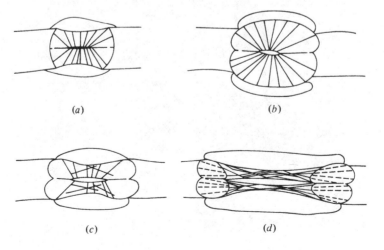

(a)   (b)

(c)   (d)

microperfusion chambers, young guard cells go through almost all stages of differentiation described above. Very young cells just after division of the guard mother cells will initiate or continue pad-like wall thickenings. The cells then enter the kidney-shaped transient and open pores. Cells in the transient will reconstrict and deposit thick walls. Those that have just passed the transient at the time culture began will continue to differentiate and lay down thick walls. Birefringence of the wall approaches the level in cells which mature on the whole plant (Fig. 9b). Time-course measurements show that it takes 1–2 h for a cell just prior to the transient to achieve full kidney shape, and 1–2 h more to reconstrict. It takes approximately 6 h for the first measurable birefringence to appear after the transient, though accelerated wall deposition undoubtedly begins earlier.

These results tell us that the kidney-like transient *is* a specific stage in guard cell differentiation which immediately precedes a rapid acceleration of wall deposition. That the transient or events immediately following it play an active role in the shift in microfibril orientation can be deduced from another observation. Those guard cells which were in the transient or younger stages at the time culture began develop thickened walls, but the cellulose in that wall often remains in the juvenile, radial alignment (Fig. 9a). Examination of these same cells by electron microscopy after fixation and sectioning shows that their microtubules are also radially aligned (Fig. 9a). Those cells in the same file which had just begun to reconstrict (but as yet had

Fig. 8. (a). Localization of potassium in maize midtransient guard cells using the cobaltinitrite procedure. ×1600. (b). Intense neutral red coloration of vacuoles in wheat guard cells at the transient stage. ×1600.

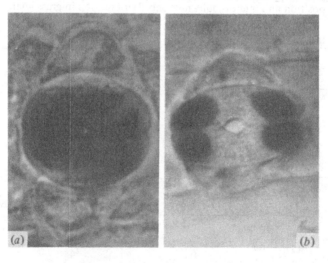

no detectable birefringence) at the time culture began also go on to lay down thick walls, but these contain normal, axially-oriented microfibrils. Thin section analysis of these cells shows that microtubules are also axial. We do not know as yet what causes this anomaly; however, it is very helpful because it emphasizes that events determining microtubule and cellulose orientation in these cells *can* be uncoupled from wall synthesis *per se*, that the uncoupling occurs around the shape transient and, by deduction, the mechanisms which determine the shift in the orientation become operative during this time. Once these mechanisms are activated and reorientation is determined, reversal does not occur. These observations tell us that we often do not completely mimic the environment of the intact seedling in our culture chambers, so that certain signals or events operative at the time of the

Fig. 9. (*a*). A cultured guard cell of timothy. This cell had not yet entered the transient stage when culture began, but after 24 h it had differentiated and deposited a thickened wall. Note, however, that microfibrils and microtubules typically remain in the radial orientation in these cells. ×34000. (*b*). Polarizing micrograph of guard cells which had reconstricted but not yet thickened their walls when culture began. After 24 h, strong wall birefringence has appeared. The microfibrils and microtubules in such cells are axial. ×1200.

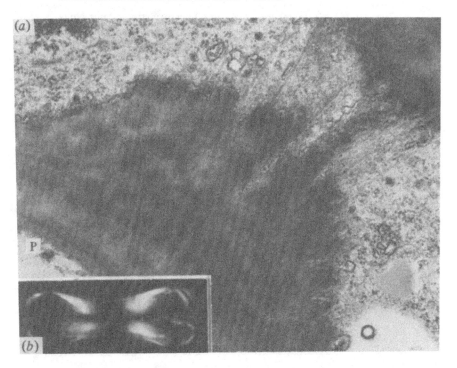

transient stage do not occur, and cortical microtubules remain in their juvenile orientation. The *amount* of wall formation appears to follow a normal course, but its orientation is anomalous. Thus, the reorientation of microtubules following the transient appears to be a specific, determined event.

That the kidney-like transient and osmotic events are important in determining wall architecture of the guard cells can be further ascertained by examining the stomatal complexes in other species. In the Cyperaceae, a family with characteristics similar to those of the Gramineae, the stomatal complexes are similar in morphology to those of the grasses and consist of two constricted guard cells each flanked by a single subsidiary cell (Ziegenspeck, 1944). The divisions leading to the formation of these cells are identical to those in the grasses. A swelling associated with the initial opening of the stoma also occurs in guard cells of *Cyperus* and *Carex* (Fig. 10*a*). However, the cells reconstrict very slowly and not nearly to the same extent as in the grasses. An accelerated thickening of the paradermal walls occurs after the transient stage (Fig. 10*a*), but often it seems to progress much more rapidly than in timothy-grass, and can become extensive *before* reconstriction is completed. When wall birefringence is analysed, it can be seen that wall microfibrils often remain *radially* oriented during this acceler-

Fig. 10. (*a*). Stomatal complexes in laboratory-stressed nut sedge (*Cyperus esculentus*). Note the strong increase in guard cell birefringence (compensated) across just 3 complexes. Features identical to those in this figure and in Fig. 10 (*b*) were found in field-grown plants. Micrograph by M. Mishkind & B. Palevitz. ×1300. (*b*). Microfibrils in such guard cells are radial as judged by the blue 'addition' colour in the mid-zone in polarized light when the Red I plate major axis is oriented as indicated (bar). ×600. (*c*). In cells grown under optimal greenhouse conditions, microfibrils are axial as judged by the yellow 'subtraction' colour. ×600.

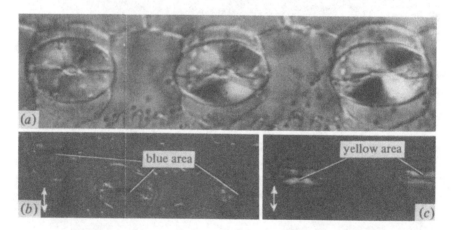

ated stage of wall deposition (Fig. 10*b*), in contrast to the axially oriented cellulose found in maturing grass guard cells. Cortical microtubules are also radial. Some axial cellulose may eventually be deposited as the cells age, and in these cells some microtubules are now axial as well. Thus, in these species which exhibit limited reconstriction, a shift to axial microtubules and cellulose may be retarded or may never occur.

These observations were made on material collected from field-grown plants. If *Cyperus esculentus* L. (nut sedge) grown under optimal greenhouse conditions is examined, a quite different finding is obtained: the pattern of wall formation is now nearly identical to that of grasses. Following the kidney-like transient, the cells reconstrict to a greater extent before significant wall thickening is observed, and wall microfibrils are now axially aligned (Fig. 10*c*). It would appear then that differences in growth conditions (e.g. water stress?) may affect critical events which determine wall architecture in guard cells. Thus we were not surprised to find that when

Fig. 11. (*a*). Very young guard or guard mother cells in timothy leaves treated with 5×10^{-3} M colchicine or 5×10^{-4} M IPC (isopropyl-*N*-phenylcarbamate) develop wall thickenings abnormally localized in the cell (IPC, 48 h) ×1800. (*b*). Guard cells treated later on develop patchy birefringence indicative of aberrant microfibril orientation, but the *site* of thickening in the mid-zone is less affected. (IPC, 36 h). ×1300.

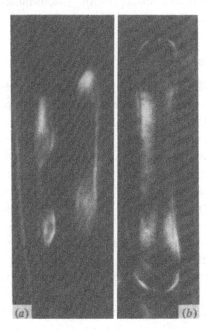

greenhouse-grown *Cyperus* plants were returned to stress conditions in the laboratory (infrequent watering, low illumination), the pattern of thickening and the orientation of cellulose in new, maturing cells reverted to that seen in field-grown plants (Fig. 10a, b). It would be of interest to know whether mature guard cells with axial cellulose behave similarly to those with radial cellulose during stomatal opening.

Microtubules are implicated in wall deposition using standard pharmacological treatments as well. As stated previously, colchicine strongly affects wall deposition in guard cells of *Allium*. We have found that colchicine and isopropyl-*N*-phenylcarbamate (IPC) have an essentially two-phase effect on wall formation and cell shaping in grass guard cells. When very young guard or guard mother cells are presented with either drug, they undergo grossly abnormal wall deposition. Large tufts of wall accumulate at various locations in the cell (Fig. 11a). The cells are abnormally shaped as well. When cells which are further along in wall deposition are so treated, the site of wall thickening is less affected and tufts are increasingly less common. However, the orientation of cellulose in the wall remains abnormal as ascertained by polarization microscopy (Fig. 11b). The wall microfibrils are laid down in random swirls yielding a patchy birefringence pattern. Again, in even more mature cells, this effect decreases as well until nearly mature cells seem unaffected by the drug. Thus, these drugs affect both the localization of wall thickening and the orientation of cellulose. Early in differentiation, they alter both processes. After the site of wall deposition is determined, they affect only the latter.

Stomatal cell interactions

It is implicit, in some of the work discussed above, that subsidiary cells influence shaping of the guard cells. From almost their inception grass guard cells are constricted on the side facing the subsidiary cells. Close examination of subsidiary cells around the time of the transient swelling of neighbouring guard cells shows that they seem to undergo reciprocal shape changes. When the leaves of corn seedlings are treated with cobaltinitrite, a reciprocal accumulation of reaction product in the two cell types appears around the time of the shape transient. Some reaction product is often present in very young subsidiary cells, but as the transient stage is approached, it shifts into the swelling guard cells. During reconstriction, precipitate is again localized preferentially in subsidiary cells.

These observations are further reinforced by other observations. Occasionally guard or subsidiary cells are injured during manipulation of the tissue. When subsidiary cells are injured, the guard cells bulge out abnormally; obviously a source of pressure on the lateral dorsal wall has been removed. When guard cells are killed, the subsidiary cells bulge abnormally on that side and drastically compress the remains of the guard cells. When young guard cells of *Allium* are treated with colchicine, they swell abnormally. However, when guard cells of timothy-grass are so treated, swelling is suppressed; that is, it takes longer to occur and never reaches the same extent as in *Allium*, even though colchicine induces wall aberrations in guard cells of timothy-grass and C-tumours on its roots. Again, it is as if the subsidiary cells serve as a barrier to excessive swellings of the guard cells.

A wealth of physiological data indicates that mature guard, subsidiary and other epidermal cells closely interact during ion movements involved in stomatal opening. Studies using micromanipulation and injection techniques indicate that the shape of mature guard cells is not only determined by their own osmotic properties but by mechanical influences arising from adjacent subsidiary cells as well (e.g. Edwards, Meidner & Sheriff, 1976). It is entirely plausible that these cells also interact during differentiation.

Several lines of evidence strongly support the idea that sister guard cells in grasses are in close communication during differentiation. Previous studies have shown that the peripheral regions of the common wall between grass guard cells contain large perforations that result from incomplete cytokinesis during division of guard mother cells (Kaufman *et al.*, 1970; Srivastava & Singh, 1972; Fig. 12). The plasma membranes of sister cells are continuous through these perforations. By serial sectioning, it can be seen that the wall is literally fenestrated by perforations. Our work indicates that some of these large 'holes' may form by wall-degradative processes late in

Fig. 12. Cross-section through the bulbous end region of guard cells of timothy-grass showing the perforations in the common wall. Note plastid with starch. ×27000.

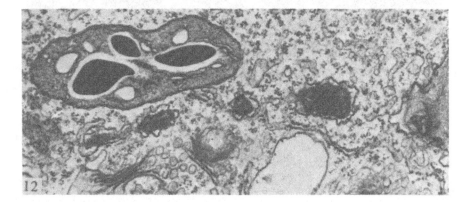

development as well. For example, microtubules and vesicles apparently unrelated to those involved in the wall thickening going on elsewhere, appear in this region as the cells mature. Myelin-like membrane whorls and multivesicular bodies thought to be associated with degradative events in other cells are also commonly seen near the wall. Thin patches occasionally seen in this part of the common wall may be areas undergoing digestion. In any event, mature guard cells seem to comprise a coenocyte in which the cytoplasm of both cells is continuous. Indeed, using conventional electron microscopy as well as stereo pairs of thick sections taken on the high voltage electron microscope, it can be seen that organelles, including plastids and elements of the endoplasmic reticulum, span these perforations. It is likely that such a free exchange of cytoplasmic materials between sister guard cells would have an impact on the course of development and stomatal function.

Most of our evidence and that of others (see Willmer & Sexton, 1979) indicates that mature guard cells are not linked to neighbouring cells *via* plasmodesmata. Although these structures are present in the walls of young

Fig. 13. (*a*). Stomatal complexes soon after division of guard mother cells in a maize leaf stained with aniline blue. The common wall exhibits the fluorescence indicative of β 1–3 glucans. (*b*). Ammonia-enhanced autofluorescence indicative of phenolic acids in oat guard cell walls. ×1800.

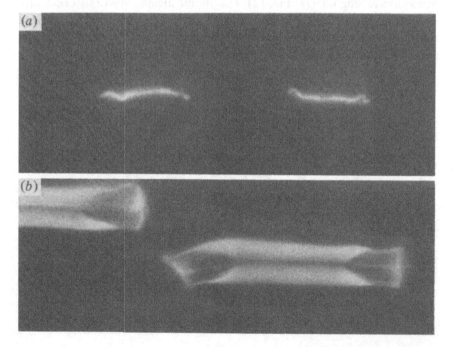

cells, they somehow disappear during development. This process may set the stage for more thorough regulation of the flow of molecules into and out of mature guard cells at the plasma membrane level.

Other wall components

Relatively little is known about the composition of guard cell walls. It is unclear whether the microfibrils which we see and which presumably control the mechanical properties of the guard cell wall consist only of cellulose or contain other polysaccharides such as pectins, mixed β-glucans and xylans. It is of interest that fluorescence staining with aniline blue indicates that the common wall between grass guard cells contains significant amounts of β1–3 glucans early in development, but much less is detectable after the transient stage when accelerated wall thickening occurs (Fig. 13a).

Molecules other than carbohydrates are present in the walls of plant cells and these may have a significant influence on the mechanical properties of the wall, its susceptibility to lysis and its permeability to water and ions etc. Stomatal cells also possess such materials. It has been traditionally accepted that the walls of grass guard cells are significantly lignified according to several criteria such as (a) the strong absorbance of the walls at 360 nm and their fluorescence emission around 450 nm, (b) the heavy deposition of MnO_2 in the wall thickening after $KMnO_4$ fixation, and (c) a similar localization of peroxidase (the enzyme known to catalyze the polymerization of phenylpropane units into lignin) using diaminobenzidine–osmium staining. Closer examination, however, indicates that only part (and perhaps only a small part) of the phenolics present in the wall constitute lignin and that the major fraction appears to be phenolic acids possibly esterified to wall polysaccharides. For example, the spectral characteristics of the ultra violet induced fluorescence are markedly altered by treatment with 0.1N ammonium hydroxide at pH 10.3. The fluorescence shifts from blue to blue-green and intensifies greatly (Fig. 13b). Spectral analyses with a micro-spectrophotometer shows that there is an approximately 25 nm shift towards longer wavelength in the main emission peak and a tenfold increase in intensity. These changes are readily reversed by re-exposure to buffer at low pH. These observations have now been made on eight grass species. It is thought that lignin does not exhibit these spectral changes in response to ammonium hydroxide, and we have found that lignified tracheal elements are less affected by ammonia. Instead it is believed that these fluorescence shifts typify esters of ferulic, diferulic and p-coumaric acids (see Hartley & Jones, 1977 & references therein). It is known that such esters are released from grass cell walls by treatment with cellulase preparations. In support of

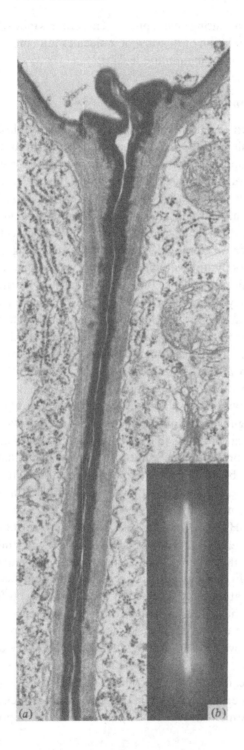

(a)   (b)

their presence in grass guard cell walls, we have found that treatment with 1 N NaOH at 20 °C for 1–2 h completely removes fluorescence from the guard cell walls of oats, but does not do so in the walls of tracheal elements Similar results were obtained for timothy-grass. Such treatment is thought to release phenolic acids selectively from carbohydrate esters (Hartley & Jones, 1977). The presence of such esters is of interest for several reasons. They may determine mechanical characteristics of the wall by their association with polysaccharides such as cellulose, and they may also serve as precursors available for future wall modifications or as defence against invading pathogens. A predominance of esterified phenolic acids might also explain the weak phloroglucinol reaction in these guard cells. A further understanding of these esters may be gained by studying their accumulation during differentiation using ammonia enhancement and fluorescence microscopy. Phenolic deposition is exactly keyed to wall thickening after the transient stage, and the appearance of fluorescence closely matches that of birefringence. Fluorescence also appears in the walls of cultured guard cells. The same wall fluorescence has been found in the guard cells of *Cyperus* and *Carex* but is absent from *Allium*.

During the initial opening of the pore, a layer of cuticle must form over the newly exposed wall surface. In grasses and *Allium* the pore cuticle is weakly autofluorescent, but can be made more fluorescent by staining with Auramine O (0.01%, pH 7.2; (Fig. 14*b*). That this layer represents a cuticle is confirmed by electron microscopy (Fig. 14*a*). In addition, the Auramine O staining layer is removed by incubation of fresh timothy-grass leaves in a cutinase preparation supplied to us by Dr P. E. Kolattukudy. Using Auramine O, we found that a thin cuticle appears *very rapidly* around the new pore in timothy-grass in the mid-transient stage and deposition stays in place as the pore elongates. Electron microscopy confirms that the cuticle is deposited rapidly after the pore opens, first as thin patches, which then merge into a uniform electron-dense layer on the surface of the wall lining the pore and in the stomatal ledges (Fig. 14*a*). The deep electron density of this cuticle distinguishes it from that covering other epidermal cells, and a transition in cuticle density can be seen at the boundary between guard and subsidiary cells. The electron density is the result of enhanced binding of osmium and is not due to heavy metals. Interestingly, only limited cuticle formation seems to occur on the walls of guard cells facing the substomatal chamber.

Fig. 14. (*a*). An electron-dense cuticle lines the pore and ledges of guard cells of timothy (cross-section). ×30000. (*b*). Auramine O-cuticle fluorescence (yellow) around the stomatal pore in oats. ×1600.

Vistas

Because of large scale wall deposition and a very active Golgi system, guard cells provide a potentially useful system to study processes of membrane flow during exo- and endocytosis. High voltage electron microscopy analysis of thick sections shows that the cortical cytoplasm contains a complex system of vesicles and membranous elements. We have recently examined lysed protoplasts obtained from *Allium* guard cells after negative staining and have found that large numbers of coated vesicles are also present near the plasma membrane. These vesicles have now been implicated in endocytosis and hormone internalization in animal cells.

Guard cells provide an ideal system for studying the morphogenetic factors and cell structures that control the deployment of microtubules. Ultimately it is this process that is of interest in cell-shape determination. Using high voltage electron microscopy, we have found small aster-like clusters of vesicles and microtubules in the cortex of grass guard cells that may serve as microtubule organizing centres. These are similar to the structures recently reported by Gunning, Hardham & Hughes (1978), and may provide an opportunity to find out more about the control of microtubule polymerization in plant cells.

We need to learn more about the mechanisms, presumably including controlled wall lysis, that govern the creation of the pore and the stomatal ledges (Stevens & Martin, 1978). We also need to know more about the larger organelles in stomatal cells. For example, a greater knowledge of the structure and composition of guard cell vacuoles is desirable, especially in light of their characteristic autofluorescence in some species (Zeiger, this volume). Guard and subsidiary cells contain many mitochondria. They also contain plastids. Those in subsidiary cells are rather nondescript; however, the structure of guard cell plastids varies widely (Allaway & Milthorpe, 1976). In some species, the plastids are prominent; in others, they are rather unimpressive. Some plastids in guard cells contain numerous starch grains (Fig. 12); others have none. In some species, internal membranes are fairly well developed (though less so than in mesophyll plastids). In others, the plastids are poorly organized, with few grana, though they do exhibit red autofluorescence. We need to know more about the control of plastid formation and its relationship to carbon metabolism in these cells. We must also improve our understanding of the role of microbodies in the carbon balance of guard and subsidiary cells. Ultimately, it is hoped that a more complete picture of all stomatal cell components will be obtained.

This research was supported by funds from the National Science Foundation and the University of Georgia.

References

Allaway, W. G. & Milthorpe, F. L. (1976). Structure and functioning of stomata. In *Water Deficits and Plant Growth,* vol. IV, ed. T. T. Kozlowski, pp. 57–102. New York: Academic Press.

Edwards, M., Meidner, H. & Sheriff, D. W. (1976). Direct measurements of turgor pressure potentials of guard cells. 2. *Journal of Experimental Botany* **27,** 163–71.

Flint, L. H. & Moreland, C. F. (1946). A study of the stomate in sugarcane. *American Journal of Botany* **33,** 80–92.

French, J. & Paolillo, D. J. (1975). The effect of the calyptra on the plane of guard cell mother cell division in *Funaria* and *Physcomitrium* capsules. *Annals of Botany* **39,** 233–6.

Gunning, B. E. S., Hardham, A. R. & Hughes, J. E. (1978). Evidence for initiation of microtubules in discrete regions of the cell cortex in *Azolla* root-tip cells. *Planta* **143,** 161–79.

Hartley, R. D. & Jones, E. C. (1977). Phenolic components and degradability of cell walls of grass and legume species. *Phytochemistry* **16,** 1531–4.

Hepler, P. K. & Palevitz, B. A. (1974). Microtubules and microfilaments. *Annual Reviews of Plant Physiology* **25,** 309–62.

Kaufman, P. P., Petering, L. B., Yocum, C. S. & Baic, D. (1970). Ultrastructural studies on stomatal development in internodes of *Avena sativa. American Journal of Botany* **57,** 33–49.

Palevitz, B. A. (1980). Comparative effects of phalloidin and cytochalasin B on motility and morphogenesis in *Allium. Canadian Journal of Botany,* **58,** 773–85.

Palevitz, B. A. & Alones, V. E. (1977). Microtubules, wall formation and a shape transient in grass guard cells. *Journal of Cell Biology* **75,** 287*a*.

Palevitz, B. A. & Hepler, P. K. (1974). The control of the plane of division during stomatal differentiation in *Allium. Chromosoma* **46,** 297–326.

Palevitz, B. A. & Hepler, P. K. (1976). Cellulose microfibril orientation and cell shaping in developing guard cells of *Allium:* the role of microtubules and ion accumulation. *Planta* **132,** 71–93.

Singh, A. P. & Srivastava, L. M. (1973). The fine structure of pea stomata. *Protoplasma* **76,** 61–82.

Srivastava, L. & Singh, A. P. (1972). Stomatal structure in corn leaves. *Journal of Ultrastructural Research* **39,** 345–63.

Stevens, R. A. & Martin, E. S. (1978). Structural and functional aspects of stomata. *Planta* **142,** 307–16.

Willmer, C. M. & Sexton, R. (1979). Stomata and plasmodesmata. *Protoplasma* **100,** 113–24.

Ziegenspeck, H. (1944). Vergleichende Untersuchung der Entwicklung der Spaltöffnungen von Monokotyledonen und Dikotyledonen in Lichte der Polariskopie und Dichroskopie. *Protoplasma* **38,** 197–224.

Ziegler, H., Shmueli, E. & Lange, G. (1974). Structure and function of the stomata of *Zea mays. Cytobiologie* **9,** 162–8.

H. MEIDNER

Measurements of stomatal aperture and responses to stimuli

Introduction

For any investigation into stomatal mechanisms or their control, it is essential to be able to measure or to estimate stomatal apertures and changes that occur in them. There are two approaches possible – microscopic measurements and estimates based on conductance methods. Although not easy and certainly time-consuming, microscopic measurements alone give precise information about the dimensions of stomatal pores and the guard cells enclosing them, and certain aspects of stomatal mechanics including the *Spannungsphase* (Stålfelt, 1963; Edwards, Meidner & Sheriff, 1976; Meidner & Bannister, 1979) could not have been recognized without these (cf. Fig. 1).

Conductance methods provide data which must be transformed to give

Fig. 1. Time course of changes in widths of stomatal pores and of the whole stomatal complex of *Vicia faba* L. Note that between 24 and 4 h the width of the stomatal apparatus changed without a measurable pore becoming visible (Spannungsphase). Rhythm in stomatal aperture, filled circles; total width of guard cells, open circles. From Stålfelt (1963).

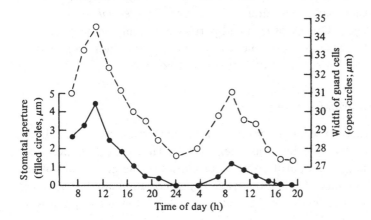

estimates of stomatal pore widths, or of epidermal and leaf conductance. These methods have contributed immeasurably to our knowledge of stomatal physiology.

Microscopic measurements

For microscopic measurements the nature of the material chosen and its subsequent treatment are both critical.

Intact leaves

In incident light, measurements may be difficult because of reflection from differently inclined waxy cell surfaces (Müller, 1872), while with moderately thick leaves transmitted light may not provide a clear image. Once a clear image has been obtained measurements must be made on about 20 stomata in order to arrive at a valid mean. During the time required for these measurements to be made, changes in aperture may occur due only to the investigation. This is most likely if the leaf is kept in air or submerged in water on the microscope stage. Immersion of the leaf, or usually a small cut-off portion of leaf, in liquid paraffin (sp. gr. 0.86) (Stålfelt, 1959) provides a clearer image and reduces the influence of changes in carbon-dioxide concentration and loss of water vapour on stomatal aperture during the measurements.

Stomatal studies of intact leaves (preferably still attached to the plant) provide information on stomatal responses to stimuli without serious disturbance of the physiological interconnections between the stomatal cells and the rest of the leaf. When combined with other methods, results from such studies are of value because they can indicate any changes in stomatal responses that occur when material is used in which the interplay between different tissues has been altered (compare *Paradermal sections* and *Epidermal strips*). On the other hand particular treatments of guard cells alone are of course not possible with such material; nor can tests of hypotheses concerning guard cell metabolism be made.

Paradermal sections

Paradermal sections containing isolated stomata mounted in liquid paraffin have been used by Mouravieff (1955). They provide material for measuring changes in aperture due to responses of the guard cells alone, because the influence of neighbouring cells has been eliminated. Mouravieff (1956) showed that many of the factors to which stomata respond were indeed effective when isolated stomata were exposed to them, i.e. processes which occur in adjacent tissues were not *essential* for guard cell responses to these factors.

Lloyd's strips

Lloyd's method (1908) of 'killing' or 'fixing' an epidermal strip immediately after taking it by plunging it into absolute alcohol, has yielded results which on analysis have proved to be unreliable. The fixed material must be mounted *via* clove oil in a suitable medium for measurements. Depending on the species both the fixing and the transfer to a mounting medium may cause uncontrolled changes in aperture. Except for field surveys, Lloyd's method cannot be recommended, cf. Table 1.

Epidermal strips

Epidermal strips give clear microscopic pictures but the value of measurements made on such material is greatly influenced by the subsequent treatment of the strips. This is important both in surveys of the state of stomata responding to different environmental conditions and in experimental work.

When strips are used it must be ascertained to what extent stripping has damaged epidermal cells. The elimination of 'tissue pressure' (Edwards & Meidner, 1979), specifically the turgor pressure of subsidiary or neighbouring cells, can profoundly affect stomatal apertures. Taking the strips in air of unknown and variable moisture content will affect stomatal aperture as uncontrolled turgor losses will occur; taking strips with open stomata while the material is submerged in water will cause reductions in stomatal pore widths as overall turgor will be increased (cf. Fig. 2; Table 2). The most reliable method appears to be to take the strips with the material submerged in liquid paraffin. Similar considerations apply to the mounting medium for epidermal strips in preparation for measurements; when mounted in water, reductions in the aperture of open stomata are likely to occur if the majority

Table 1. *Changes in stomatal pore widths of* Zebrina pendula *Schnizl. after fixing by Lloyd's method (1908)*

Means of 100 measurements	
intact leaf (μm)	Lloyd's strip (μm)
13.4	11.4
9.6	8.8

of epidermal cells remained undamaged; while in strips with a sub-stantial number of epidermal cells ruptured apertures may increase as guard cells alone gain in turgor. Once more, mounting the strips in liquid paraffin would seem to be the most reliable procedure. Covering epidermal strips for prolonged periods with a cover-glass may produce anaerobic conditions that may affect guard cell metabolism and thus lead to changes in aperture.

Table 2. *Effect of epidermal turgor on stomatal pore widths in* Tradescantia virginiana L. *Means of eight measurements. From Meidner & Bannister (1979)*

At zero epidermal turgor (μm)	At maximum epidermal turgor (μm)
4.8	2.5
8.0	3.6
15.0	8.0

Epidermal strips have been used extensively for experimental investi-gations concerned with hormone influences, ion movements, metabolic changes in guard cells, pressure measurements and electro-physiological properties. For some of these investigations, epidermal strips are the most suitable material, provided that the special conditions associated with the treatment of the tissues are kept in mind. Nevertheless, the manipulation of

Fig. 2. Stomata in a single epidermal strip taken from an illuminated leaf of *Tradescantia virginiana* L. Half the strip was mounted in liquid paraffin (left) while the other half was mounted in water (right). Reduction in pore widths occurred because the epidermal cells gained full turgor when water became available.

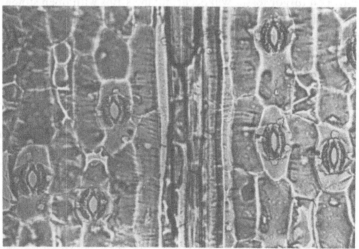

the strips may influence stomatal responses to subsequent experimental treatments. Exposure to media of low pH (pH = 4.5) kills most of the epidermal cells within about two hours, but appears not to affect guard cell properties in any significant way. Treatments at pH values below 4.5, however, are known to affect severely the properties of guard cells (Marrè *et al.*, 1974). Such effects become clearly recognizable in treatments exceeding 2 h but they must begin to be of consequence right from the start of the treatment. Both metabolic and wall properties can be affected. Provided one is aware of these effects, such treatments yield uniform material for experimental work designed specifically to test responses of guard cells not in contact with live neighbouring cells. This is a limitation from the point of view of translocation of metabolites between cells, although changing the bathing solution allows studies of the import and export of guard cell solutes.

If hypotheses are to be tested which concern guard cell metabolism, epidermal strips with all but the guard cells killed by treatment at pH = 4.5 (Squire & Mansfield, 1972) provide the best material obtained in the most elegant and labour-saving way. The choice of the pH buffer system is important as many of these buffer systems contain potassium or other metallic ions and weak organic acid radicals which may well be involved in guard cell metabolism and could thus confound results obtained. The 'Good'-buffers in the relevant ranges (Good *et al.*, 1966) must be treated with caution. They are imperfectly buffered against carbon dioxide and they affect the metabolism of guard cells if used for prolonged periods. This is most readily demonstrated by the very unusual neutral-red staining of epidermal tissue treated with these buffers. Whereas guard cells usually accumulate the stain strongly and the inner subsidiary cells remain unstained, in strips treated with MES it is the inner subsidiary cells which stain (weakly) while in the guard cells the staining is noticeably reduced.

Microscopic examination of the degree of stomatal opening in epidermal strips has elucidated important aspects of stomatal physiology into which porometry could not give an insight. Although in epidermal strips the naturally maintained water relations of the epidermal cells are severely disturbed and metabolite translocation between epidermal and mesophyll tissue is prevented, valid conclusions have been arrived at concerning potassium transport into guard cells (Imamura, 1943; Fujino, 1967; Fischer, 1968), the potassium shuttle in maize stomata (Raschke & Fellows, 1971), the involvement of malate metabolism (Willmer, Pallas & Black, 1973) bio-assays for ABA and other hormone effects (Ogunkanmi, Tucker & Mansfield, 1973), humidity effects (Lange *et al.*, 1971), pressure relations of stomatal cells (Meidner & Edwards, 1975) and certain organelle changes associated with stomatal movements (Heller & Resch, 1967; Humbert & Guyot, 1972).

The following two techniques are mentioned here because they permit correlation between guard cell contents and stomatal aperture. By themselves they are not suitable for the measurement of changes in stomatal aperture.

Micro-surgically excised guard cells

This very specialized technique is being pioneered by Outlaw (Outlaw & Lowry, 1977). To be certain about metabolic changes occurring in guard cells in response to stimuli, measurements must be made on these cells alone, completely isolated from all other cells so that translocation is made impossible. Absolute values of changes in guard cell starch content have been obtained with this technique (Outlaw & Manchester, 1979) whereas hitherto only qualitative starch scores (Williams & Barrett, 1954) could be estimated.

Guard cell protoplasts

This technique represents the ultimate in dealing with the metabolically 'live' part of stomata but at the cost of dispensing with the cell wall. The latter is known to influence, for instance, electro-potentials across the plasmalemma (Moody & Zeiger, 1978; Kinnersley, Racusen & Galston, 1978) and is more than likely to have other important effects on metabolic processes. None the less, effects of hormones on guard cell protoplasts have been observed (Schnabl, Bornman & Ziegler, 1978) and a stimulus of blue light on guard cells (Zeiger & Hepler, 1977) has been identified by the microscopic observation of the swelling and shrinking of naked guard cell protoplasts.

Micro-relief methods

These date from 1901 (Buscaloni & Polacci) and were revived by Sampson (1961). Micro-relief peels are a convenient permanent record of the state of stomatal opening and are readily examined and evaluated microscopically. They are more suitable for some leaves than for others as the degree of entry of the plastic matrix into the stomatal pore depends on a variety of morphological features. Only in a few species can the matrix enter the stomatal throat, which acts as the controlling resistance (Gloser, 1967). For this reason micro-relief peels can fail to show changes in the relevant pore dimensions. As a rule the distance between the two opposing guard cell ridges may remain relatively unchanged when in fact the throat of the pore may have changed considerably; in short, guard cell ridges do not change their position proportionately with guard cell ventral walls (cf. Fig. 3). This emphasizes that guard cell deformation does not only affect cell width in the direction parallel to the leaf surface but also cell depth in the

Fig. 3. Wide-open and half-open stomata of *Tradescantia virginiana* L.: (*a*) with the ventral walls forming the throat of the pore in focus and (*b*) with the guard cell ridges in focus. While the pore widths changed from 9 to 18 μm (2:1), the distance between the ridges changed from 18 to 24 μm (1.3:1).

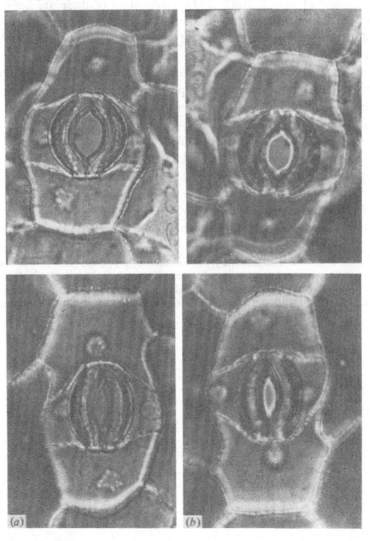

(*a*)   (*b*)

direction at right angles to the leaf surface (cf. Fig. 4). This method has not contributed markedly to the study of stomatal physiology and is at best a convenient survey tool similar in value to infiltration methods (compare studies on *Intact leaves*, p. 26).

Conductance methods

Liquids by infiltration

Many prescriptions for infiltration liquids have been proposed and used (Molisch, 1912; Hack, 1974). Almost all of these are known to be harmful to leaf tissue. Alvim & Havis claim that mixtures of 'Nujol' and *n*-dodecane are not injurious so that they can be used repeatedly in the same region of the leaf blade. Several procedures are available for the estimation of stomatal aperture from infiltration studies. (i) Measurement of *rates of entry* of the liquids, i.e. the disappearance of standard size droplets. (ii) Measurement of the *degree of entry* by the spread of the liquids in the mesophyll tissue. (iii) *Entry or non-entry* depending on the surface tension and/or viscosity of the applied liquids.

In all procedures the size of the droplets must be kept constant. Quantitative estimates of stomatal conductance can best be obtained with a graded series of liquids of different surface tensions and/or viscosities (cf. Wormer & Ochs, 1959; Alvim & Havis, 1954).

One basically different approach was made by Froeschel and Chapman

Fig. 4. Diagrammatic representation of proposed dimensional changes occurring during stomatal opening showing the relatively small change in the distance between the guard cell ridges and an assumed change in guard cell depth which would explain these dimensional relationships.

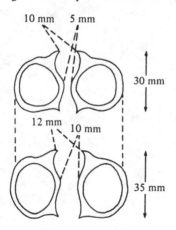

(1951) who used water as the infiltration liquid and measured the pressures required to force water into the leaf. Since infiltration methods can only be considered as semi-quantitative field methods, the complication of applying pressures to force the liquid into the leaf would seem to be unwarranted except perhaps for such leaves as the conifer needle (Fry & Walker, 1967).

Several theoretical considerations have been applied to the interpretation of results obtained by infiltration methods, concerning the contact angles of the liquids on the leaf surface (Fogg, 1948; Schönherr & Bukovac, 1972) and changes in the slopes of the outer walls of the guard cells at different pore widths (Müller, 1872). These factors affect the validity and interpretation of results obtained by infiltration methods. Infiltration scores have been correlated with stomatal aperture in tobacco (*Nicotiana tabacum* L.) (MacDowall, 1963) and with mass-flow resistance in cotton (*Gossypium hirsutum* L.) (Moreshet & Stanhill, 1965).

Taking all of their shortcomings into account, it must be concluded that infiltration methods cannot provide reliable and precise information and are suitable only for approximate measurements and for field survey work if better procedures are not practicable. A thorough study of infiltration methods has been made by Hack (1974, 1978).

Gases by porometry

Porometer techniques are based on the passage of gases through the leaf or of water vapour out of the leaf, either by diffusion or down an artificially applied pressure gradient. Measuring procedures depend on the pressures required to produce a mass flow of a gas mixture, usually air, i.e. *mass flow porometers*, or on the rate of diffusion of a gas, often nitrous oxide, hydrogen, or helium, through the leaf, i.e. *diffusion porometers,* or on the rate of diffusion of water vapour out of the leaf into a known external atmospheric humidity, i.e. *hygroscopic indicators* and *transpiration porometers*.

For these measurements a great variety of apparatus and devices has been developed; the main types are dealt with below and their most appropriate uses are discussed. Porometric investigations do not destroy the leaves although they may temporarily alter the leaf functions. The main advantage of porometers is that they provide quantitative measurements of the collective conductances of several thousand stomata together. These measurements are, however, confounded to some degree by other resistances contributing to the total leaf conductance because the flow paths for gases or water vapour include, in addition to the stomatal pores, other pathway sections with changeable conductances (cf. Fig 5). Therefore, porometer measurements do not measure stomatal conductance within the leaf attachment clamp alone.

Mass flow

Viscous flow porometers date from 1867 (Müller, 1869) and became more widely used following Darwin & Pertz's (1911) design. They were variously modified in a number of ways.

Airflow systems. A flow of air can be brought about by suction or pressure from aspirators, preferably designed to give constant pressure differences across the leaf of air of known humidity. The latter property was found to be of importance when leaf resistance was estimated by comparison with fixed capillary resistances because air drawn through a leaf becomes humid, so that the air should be dried before it flows through the capillary resistances (Knight, 1915; Gregory & Pearse, 1934; Heath & Russell, 1951; Spanner & Heath, 1951).

Other methods of creating an airflow through a leaf include the use of a column of water (Darwin & Pertz, 1911), or a drop of mercury descending down a capillary glass tube (Maskell, 1928; Meidner, 1955; Weatherley, 1966), the inflation and deflation of a rubber bulb (Alvim, 1965; Meidner, 1965 *a*), or the withdrawal of a syringe plunger (Milburn, 1979).

Fig. 5. Diagrammatic representation of paths for mass flow of air in a leaf with a porometer cup attached.

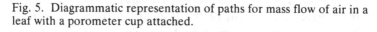
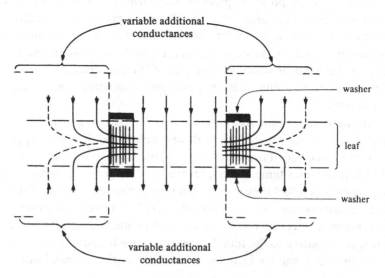

Measuring systems. In most of the earlier porometers and some of the simpler later models, the measurements were based on the time required to let a predetermined volume of air pass through the leaf. Darwin & Pertz (1911) had already used this method and innovations concerned minor details only. A different measuring system was introduced by Gregory & Pearse (1934) with their resistance porometer, in which the pressure drop across a leaf was quantitatively determined in relation to calibrated fixed resistances using manometers as the measuring devices. For this instrument the 'Gregory and Pearse resistance unit' was introduced; it had the value 3.77×10^8 cm^{-3} (Heath & Russell, 1951), which can be used to convert these units to absolute resistance values. Recording resistance porometers with multiple leaf attachment cups (Heath & Mansfield, 1962) made experiments less cumbersome and demonstrated the uniformity of stomatal movements in different plants kept under similar conditions.

A refinement of the resistance porometer is the Wheatstone bridge porometer with its differential null-reading manometer combined with a fine needle valve resistance used to balance the leaf resistance. The Wheatstone bridge porometer is probably the most sensitive mass flow porometer.

Another measuring system was employed by Sheriff & McGruddy (1976) in their hot wire constant pressure porometer. In this instrument the air passing through the leaf cools a fine filament, thereby changing its resistance. Since the filament forms one arm of an electrical Wheatstone bridge the out-of-balance bridge current can be recorded as the measure of leaf conductance.

Leaf attachment clamps. Besides the variations in the mechanical construction of the clamp, the most important development in the mode of leaf attachment was the replacement of the permanently attached clamp by detachable or ventilated clamps. This became imperative since Freuden-berger (1940) and Heath (1948) had shown that stomata in a permanently attached clamp were at different apertures from those outside it. This was the original recognition of the effect of carbon dioxide on stomata. Both carbon dioxide enrichment of the air in the cup when kept in the dark and carbon dioxide depletion of the air when the leaf was illuminated caused unrepresentative stomatal resistances to be recorded. Various designs of detachable cups (Heath & Mansfield, 1962; Meidner, 1965a) and of ventilated cups (Raschke, 1965; Turner, Pedersen & Wright, 1969) were introduced (cf. Fig. 6).

It must be pointed out that the pressure of the clamp on the leaf should remain constant as this has a bearing on the resistance to airflow offered by the mesophyll tissue (Meidner, 1955a; Bierhuizen, Slatyer & Rose, 1965).

This mesophyll resistance is not constant and constitutes a complicating factor for the interpretation of all viscous and diffusive flow porometer readings (Heath, 1959). The magnitude of this mesophyll resistance was estimated with another form of clamp, 'the double cup' (Knight, 1916) and by the use of specially formed gelatine washers which directed the airflow in

Fig. 6. (a). Detachable porometer leaf clamp (after Spanner & Heath, 1951) (b). Ventilated leaf disc clamp. After Raschke (1965).

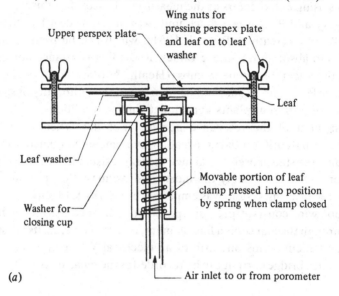

(a)

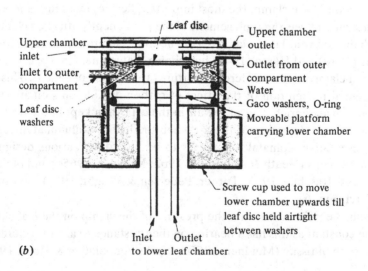

(b)

the leaf discs and annuli along certain paths (Meidner, 1955*b*). A further requirement for leaf clamps which remain attached to a leaf for any length of time is that they should be transparent.

Limitations of mass flow porometry. In practice only amphistomatous leaves and leaf discs can be used with these instruments. Except with leaf discs an unknown variable number of stomata outside the attachment clamp offer resistance to air flow depending on the state of stomatal opening. In consequence there is also within the leaf a mesophyll path length which connects the variable regions outside the clamp, *via* the width of the washer, with the path through the stomata inside the cup. These additional path resistances are difficult to assess but mass flow porometer results can be corrected for them. An analysis of these additional mesophyll path resistances has been presented by Penman (1942; Penman & Schofield, 1951). Heath (1941) used this theory to estimate stomatal resistances from the measured total leaf resistances. The use of leaf discs overcomes these difficulties provided the cut edges are properly sealed (Raschke, 1965; Glinka & Meidner, 1968).

When using amphistomatous leaves for viscous flow porometry it is essential to determine stomatal sizes and densities in the two epidermes and to relate the measured resistances to that epidermis which offers the higher resistance. However this may not be a function of size and density alone because stomatal movements in the two epidermes are not necessarily synchronous. Mass flow porometry cannot adequately differentiate between stomatal resistances in the two epidermes.

There are some amphistomatous leaves which are heterobaric (Meidner,

Fig. 7. Changes in the intercellular mesophyll resistance of two leaf species. From Meidner (1955*b*).

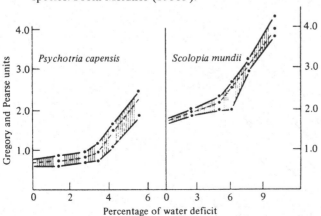

1955 *a*; Mansfield & Heath, 1963) and some of these have equal numbers of stomata in the two epidermes. Such leaves are unquestionably the most suitable material for mass flow porometry as the interference by additional variable path resistances with the measurements is negligible and the resistance of the two epidermes is often the same.

In spite of these complicating factors, mass flow porometry is an appropriate, non-destructive method for detecting stomatal responses to experimental conditions in intact leaves. If suitable leaves are used and the measurements are adequately corrected the results provide not only valid records of stomatal movements in response to stimuli but also an insight into interactions between stimuli in their effects on stomata.

Particularly useful information has been obtained concerning the effects of light intensity and quality, external and internal carbon dioxide concentration, temperature, water stress, oscillations and endogenous rhythms, and indications of hormonal dependence of stomata, especially ABA and probably kinetin. Mass flow porometry has also been applied to the study of the dynamics or kinetics of stomatal movements. This is in need of further investigation and mass flow porometry is probably the most appropriate method because intercellular mesophyll conductance and changing leaf water content influence mass flow the least.

Finally, it may be noted that the relation between mass flow and diffusive flux was first estimated theoretically (Maskell, 1928) and has since been experimentally determined (Meidner & Spanner, 1959; Slatyer & Jarvis, 1966; cf. Fig. 8).

Fig. 8. The relation between mass flow and diffusive flow conductance for ellipse-shaped stomata as measured with a combination porometer for mass flow (Wheatstone bridge) and diffusive flow (differential transpiration). The slopes of lines vary between 0.46 and 0.52. From Meidner & Spanner (1959).

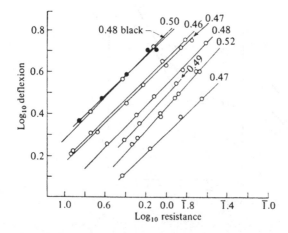

Diffusive gas flux

Since stomata contribute to the regulation of diffusive fluxes of carbon dioxide and water vapour, their diffusive conductances are perhaps of more immediate importance than their mass flow conductance. Thus, experiments which aim to correlate rates of carbon dioxide uptake and transpiration with environmental and internal factors and with leaf conductance rather than stomatal conductance *per se* gain in directness if diffusive conductance is measured without the need to transform mass flow data into diffusive conductance.

Measurements of diffusive gas flux from one epidermis to the other suffer, although perhaps less severely, from the same uncertainties as mass flow measurements because of differences in stomatal sizes and densities in the two epidermes and the complications of a variable resistance path in the mesophyll.

Basically, two types of diffusive gas flux porometers have been used invariably with amphistomatous leaves and it would probably be unrealistic to use any other.

Hydrogen porometers. Gregory & Armstrong (1936) measured the rate of electrolytically evolved hydrogen which balanced the rate of diffusion of hydrogen through the leaf. This measurement was taken as an estimate of the diffusive conductance of the leaf. The apparatus involved lengthy exposure of the leaves to hydrogen, which is harmful to them. To overcome this defect Spanner (1953) constructed the Dufour effect porometer which allowed for an estimate of leaf conductance by measuring the magnitude of the Dufour effect following a very brief release of hydrogen on one side of the leaf. These hydrogen diffusion porometers are strictly laboratory instruments and are more ingenious than useful.

Inert gas porometers. The second type of diffusive gas porometer makes use of such readily identifiable inert gases as nitrous oxide (Slatyer & Jarvis, 1966), helium (Farquhar & Raschke, 1978) or radioactive argon (Moreshet, Stanhill & Koller, 1968), which are allowed to diffuse from one side of the leaf to the other. The gas is mixed with an air stream which flows at a known volume rate over one side of the leaf that forms a membrane in a cuvette. The inert gas diffuses through the leaf into an air stream passing over the other epidermis so that by gas analysis or radioactivity counts, the rates of diffusion can be calculated and are indeed a direct measure of diffusive conductance of the leaf. The ancillary equipment is complex and this restricts these instruments to laboratory use. They are suitable, however, for simultaneous measurements of photosynthesis and transpiration and have been used in this way to clarify intercellular mesophyll resistance (Jarvis, Rose & Begg, 1967), wall resistance (Meidner, 1965b) and a postulated influence on the stomatal apparatus of rates of photosynthetic production of metabolites (Wong, Cowan & Farquhar, 1979).

Water vapour diffusion

Instead of measuring rates of diffusion of externally applied gases through intact leaves the ever-present diffusion of water vapour through the epidermes of leaves presents the most natural way of estimating leaf conductance. Even so, complicating uncertainties remain, chiefly the variable degree of saturation of the leaf airspace, especially the sub-stomatal cavity. This is increasingly acknowledged to be a major site of evaporation within the leaf, precisely because its vapour density is likely to be more variable and lower than that in the interior of the leaf.

Hygroscopic indicators

A variety of such indicators has been used since the end of last century. The Horn or Hair hygroscope (Darwin, 1898) and the cellophane method (Weber, 1927) are mentioned here more as historic curiosities than aids for estimating leaf conductance although the principle of using hygroscopic sensors is, of course, the basis of modern transpiration porometers.

Cobalt chloride papers (Stahl, 1894). These can be used to measure the time it takes for the blue (dry) colour to change to a pink shade when water vapour diffuses out of the leaf to which the papers are attached. The method was improved by Shreve (1919) and again by Henderson (1936). Although it lacks refinement it is the earliest practical vapour diffusion method of estimating stomatal conductance (Milthorpe, 1955). With careful handling it can give reliable results and has the advantage that it can be used on the two epidermes simultaneously; this cannot be done with mass flow porometry. Equally important, cobalt chloride papers are suitable for hypostomatous leaves. The method presupposes that the leaf airspace is fully saturated at all times and that cuticular vapour release is negligible and constant at all times. Measurements made at different leaf temperatures are not comparable and it would be impracticable to attempt a correction for temperature effects. As an introduction to diffusive flux methods, however, cobalt chloride papers are instructive (Bailey, Rathacher & Cummings, 1952).

The corona discharge hygrometer (Andersson & Hertz, 1955). This is a precise and sophisticated instrument which relies on the different vapour densities above a transpiring leaf with stomata at different states of opening. The apparatus is a laboratory assembly which has been used by Andersson & Rufelt (1954) to good advantage; it does not involve attaching the leaf to a clamp but, rather, enclosing it in a cuvette. There are other assemblies (see below, transpiration porometers) which rely for the estimation of leaf conductance on cuvettes in order to measure differences in water vapour densities in controlled volume flows of air before and after the air passes over the leaf.

Transpiration porometers

There are two kinds of instruments which measure the amounts of water vapour given off per unit leaf area per unit time. In the most straightforward kind the water vapour enrichment of an airstream of known moisture content passing over a leaf is measured; in the other kind, known as steady-state transpiration porometers, the air flow over the leaf is adjusted so that its vapour density remains constant at a predetermined level (Beardsell, Jarvis & Davidson, 1972). The steady state property is of great value. Both kinds of instrument require as a rule that the leaf be enclosed in a cuvette and exposed to an air stream of some magnitude. This makes them unsuitable for field measurements. The calibration of these instruments in resistance or conductance units is complex and estimates of leaf conductance are calculated from very critical measurements of rates of air flow, leaf and air temperatures and water vapour densities. To obtain the required accuracy complex control gear is needed. The calculations assume that the leaf internal airspace is saturated at the prevailing leaf temperature, an assumption which is not always justified. Nevertheless, these instruments have their applications especially for certain leaf types but they should not be considered as suitable for investigations which aim to test specifically changes in stomatal aperture occurring naturally in the field.

The differential transpiration porometer (Meidner & Spanner, 1959) does not measure transpiration rates directly but their differential cooling effects. Transpiration rates are controlled by creating different water vapour densities above two adjacent spots on one leaf surface and measuring the resulting temperature difference on the opposite surface. Thus, two small areas of leaf only are exposed to the impinging air jets of 5 km h^{-1}. Attachment of the device to the leaf is not airtight and does not restrict the leaf as does enclosure in a cuvette. The double hinged clamp probably deserves wider application as its jaws move parallel to each other. The instrument is strictly a laboratory assembly and cannot readily be calibrated in conductance units. The differential transpiration porometer is, however, one instrument which can be combined to measure diffusive and mass flow conductances practically simultaneously, thereby testing experimentally the theoretically determined relationship between diffusive flux and mass flow (cf. Fig. 8). This combination instrument has in addition provided evidence of changes in transpiration rates which were not accompanied by changing stomatal conductances as measured by mass flow porometry. Although these are perhaps not the most satisfactory records of such discrepancies between stomatal conductance and transpiration rates they indicate that changes in water vapour densities inside the leaf do occur. This can be understood more readily today than at the time (1958) since it is now recognized that cell walls bordering sub-stomatal cavities can develop very low matric potentials (cf. Fig. 9).

The remaining two types of transpiration porometer incorporate humidity sensors as the essential measuring device, allowing leaf conductance to be expressed in appropriate units of diffusivity. The condensation porometer of Moreshet & Yocum (1972) has as its measuring system the time required for condensation droplets to appear on a polished plate kept at zero degrees Celsius. Since the condensation plate is kept at constant temperature, measurements obtained with this instrument require correction only for leaf temperature. In this respect the condensation porometer is more convenient than the sensor element transpiration porometers dealt with below, but the instrument is complex and this has perhaps contributed to its not being used extensively. The original design was fitted with two cups so that both upper and lower epidermal conductances could be determined simultaneously.

Fig. 9. Changes in diffusive conductance as measured with the differential transpiration porometer while mass flow conductance remained substantially unaltered. Mass flow was measured with the Wheatstone bridge porometer. From Meidner & Spanner (1959).

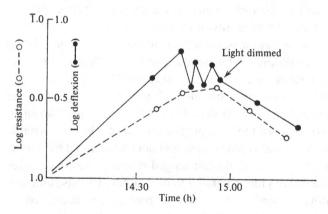

Most transpiration porometers are usually referred to as diffusion porometers and they were developed from the apparatus made originally by Wallihan (1964). They have reached a high degree of perfection and automation and are today probably the most widely used type of porometer, especially in the field, because they are able to measure both upper and lower epidermal conductances separately in quick succession or simultaneously with a double cup. The measurements are of epidermal or leaf conductance for water vapour transfer. This point must be stressed because once again the assumptions are not warranted that cuticular vapour loss is constant and negligible at all times or that the leaf air space is at all times 100% saturated with water vapour. A change in leaf conductance need not necessarily reflect a change in stomatal conductance.

The use of sensor element porometers is not as straightforward as is often assumed. The degree of dryness achieved in the attachment clamp between readings and errors in the measurement of the temperatures of both leaf and sensor are two sources of error that render suspect a considerable amount of work carried out with the early instruments. The non-steady state measurement is also open to criticism. However, some of the commercially available models claim to overcome some of these weaknesses by automatic corrections.

Attention should be paid to the attachment clamp. For amphistomatous leaves a double cup should be used as it can be shown that data collected with clamps open to the atmosphere on the side of the leaf not used for the measurement differ from data collected with both epidermes enclosed. Double cups will also give a direct measure of total leaf conductance and may demonstrate differential changes in conductances between upper and lower epidermes. Whether the upper clamps used are opaque or transparent can also be of consequence depending on the mode of use, especially the length of time the clamp remains in position. In some instruments the attachment cups are ventilated by a fan to minimize boundary layer effects (e.g. Bravdo, 1972). The pros and cons of this arrangement are debatable. For investigations aiming at measurements of stomatal movements in the field, it is probably best to minimize unusual disturbance of the air adjacent to the leaf surface and short periods of attachment of the unventilated clamp are safer than leaving the clamp on for extended periods. The presence of a vapour sink in the form of the humidity sensor some distance away from the leaf imposes of necessity an artificial environment. If the stomata in the species used are sensitive to atmospheric humidity, field measurements made with any of the transpiration meters using cuvettes or with ventilated porometers lose in validity.

The sensor element porometer is the most appropriate instrument for certain field measurements, and also for laboratory studies involving responses of stomata in the upper and lower epidermes to illumination from above and below with light of different quality and quantity.

Conclusions

Measurements of changes in stomatal aperture concentrate attention on the major function of stomata in the life of plants: the regulation of gas exchange. Thus early studies in stomatal physiology were directed towards identifying stomatal responses to external stimuli using mass flow porometers. Later, attention was paid to the dynamics of the movements of guard cells and their physiological relation to the subsidiary and other neighbouring cells. The close connection between stomata and the rest of the leaf and indeed the whole plant became a major topic of study. Circadian rhythms and oscillations due to changing plant water relations were discovered and later correlated with measurements of transpiration rates using diffusive flux methods. These studies graded over into those concerned with metabolism of guard cells, ultimately responsible for their turgor changes, for which porometry is not applicable. Microscopic measurements and biochemical treatments of epidermal strips have their place here.

With so many distinct objectives of different investigations no single method can be appropriate for all studies. Different kinds of porometry are suitable for different types of investigations and studies using epidermal strips, isolated stomata, guard cells or protoplasts provide essential complementary approaches to porometry. All these contribute to the gradually growing understanding of the workings and functions of stomata.

Considerable detail about techniques for the measurement of stomatal aperture and diffusive resistance can be found in Bulletin 809, Washington State University, College of Agriculture Research Center, 1975.

References

Alvim, P. (1965). A new type of porometer for measuring stomatal opening and its use in irrigation studies. In *Methodology of Plant Eco-physiology,* ed. F. E. Eckardt, pp. 325–29. Paris: UNESCO.

Alvim, P. & Havis, J. R. (1954). An improved infiltration series for studying stomatal opening as illustrated with coffee. *Plant Physiology* **29**, 27–8.

Andersson, N. E. & Hertz, C. H. (1955). Positive Spitzenentladung als Hygrometer. *Zeitschrift für angewandte Physik* **7**, 361–7.

Andersson, N. E. & Rufelt, H. (1954). New fast recording hygrometer for plant transpiration measurements. *Physiologia Plantarum* **7**, 753–68.

Bailey, L. F., Rathacher, J. S. & Cummings, W. H. (1952). A critical study of the cobalt chloride method of measuring transpiration. *Plant Physiology* **27**, 563–74.

Beardsell, M. F., Jarvis, P. G. & Davidson, B. (1972). A null-balance diffusion porometer suitable for use with leaves of many shapes. *Journal of Applied Ecology* **9**, 677–90.

Bierhuizen, J. F., Slatyer, R. O. & Rose, C. W. (1965). A porometer for laboratory and field operation. *Journal of Experimental Botany* **16**, 182–91.

Bravdo, B. A. (1972). Photosynthesis, transpiration, leaf stomatal and mesophyll resistance measured with a ventilated diffusion porometer, *Physiologia Plantarum* **27**, 209–15.

Buscaloni, L. & Polacci, G. (1901). L'applicazione delle pellicole di collodio allo studio di alcuni processi fisiologici delle piante ad in particolor modo pella transpirazione. *Atti Istituto botanico Universita Pavia* **2**.

Darwin, F. (1898), Observations on stomata, *Philosophical Transactions of the Royal Society of London,* Series B **190**, 531–621.

Darwin, F. & Pertz, D. F. M. (1911). On a new method of estimating the aperture of stomata. *Proceedings of the Royal Society of London,* Series B **84**, 136–54.

Edwards, M., Meidner, H. & Sheriff, D. W. (1976). Direct measurements of turgor pressure potentials of guard cells. *Journal of Experimental Botany* **27**, 163–71.

Edwards, M. & Meidner, H. (1979). Direct measurements of turgor pressure potentials IV. *Journal of Experimental Botany* **30**, 829–37.

Farquhar, G. D. & Raschke, K. (1978). On the resistance to transpiration of the sites of evaporation within the leaf. *Plant Physiology* **61**, 1000–5.

Fischer, R. A. (1968). Stomatal opening; role of K^+ uptake by guard cells. *Science* **160**, 784–5.

Fogg, G. E. (1948). Adhesion of water to the external surface of leaves. *Discussions of the Faraday Society* **3**, 162–66.

Freudenberger, H. (1940). Die Reaktion der Schliesszellen auf Kohlensäure und Sauerstoff Entzug. *Protoplasma* **35**, 15–54.

Froeschel, P. & Chapman, P. (1951). Method of measuring the size of stomatal aperture. *Cellule* **54**, 235–50.

Fry, K. E. & Wallace, R. B. (1967). A pressure infiltration method for estimating stomatal opening in conifers. *Ecology* **48**, 155–7.

Fujino, M. (1967). Role of ATP and ATPase in stomatal movements. *Science Bulletin of the Faculty of Education, Nagasaki* **18**, 1–47.

Glinka, Z. & Meidner, H. (1968). The measurement of stomatal responses to stimuli in leaves and leaf discs. *Journal of Experimental Botany* **19**, 152–66.

Gloser, J. (1967). Some problems of the determination of stomatal aperture by micro-relief method. *Biologia Plantarum* **9**, 28–33.

Good, N. E., Winget, G. G., Conolly, T. N., Izawa, S. & Singh, R. M. (1966). Hydrogen ion buffers for biological research. *Biochemistry* **5**, 467–78.

Gregory, F. G. & Armstrong, J. I. (1936). The diffusion porometer. *Proceedings of the Royal Society of London,* Series B **121**, 27–42.

Gregory, F. G. & Pearse, H. L. (1934). The resistance porometer and its application to the study of stomatal movement. *Proceedings of the Royal Society of London*, Series B **144**, 477–93.

Hack, H. R. B. (1974). The selection of an infiltration technique for estimating the degree of stomatal opening. *Annals of Botany, N.S.* **38**, 93–114.

Hack, H. R. B. (1978). Stomatal infiltration in irrigation experiments on cotton, sorghum, groundnuts and wheat. *Annals of Botany, N.S.* **42**, 509–47.

Heath, O. V. S. (1941). A critical study of the resistance porometer. *Annals of Botany, N.S.* **5**, 455–500.

Heath, O. V. S. (1948). Control of stomatal movement by a reduction in the normal carbon dioxide content of the air. *Nature* **161**, 179–81.

Heath, O. V. S. (1959). The water relations of stomatal cells. In *Plant Physiology, A Treatise,* ed. F. C. Steward, vol. II, pp. 193–50. New York: Academic Press.

Heath, O. V. S. & Mansfield, T. A. (1962). A recording porometer with detachable cups operating on four separate leaves. *Proceedings of the Royal Society of London*, Series A **156**, 1–13.

Heath, O. V. S. & Russell, J. (1951). The Wheatstone bridge porometer. *Journal of Experimental Botany* **2**, 111–16.

Heller, F. O. & Resch, A. (1967) Funktionell bedingter Strukturwechsel der Zellkerne in den Schliesszellen von *Vicia faba*. *Planta* **75**, 243–52.

Henderson, F. Y. (1936). The preparation of three colour strips for transpiration measurements. *Annals of Botany* **50**, 321–4.

Humbert, C. & Guyot, M. (1972). Modifications structurelles des cellules stomatique. *Comptes rendus Académie des Sciences, Paris* **274**, 380–2.

Imamura, S. (1943). Untersuchungen über den Mechanismus der Turgorschwankungen der Schliesszellen. *Japanese Journal of Botany* **12**, 251–346.

Jarvis, P. G., Rose, C. W. & Begg, J. E. (1967). An experimental and theoretical comparison of viscous and diffusive resistances to gas flow through amphistomatous leaves. *Agricultural Meteorology* **4**, 103–17.

Kinnersley, A. M., Racusen, R. H. & Galston, A. W. (1978). A comparison of regenerated cell walls in tobacco and cereals. *Planta* **139**, 155–41.

Knight, R. C. (1915). A convenient modification of the porometer. *New Phytologist* **14**, 212–16.

Knight, R. C. (1916). On the use of the porometer in stomatal investigations. *Annals of Botany* **30**, 59–76.

Lange, O. L., Lösch, R., Schulze, E-D. & Kappen, L. (1971). Responses of stomata to changes in humidity. *Planta* **100**, 76–86.

Lloyd, F. E. (1908). The physiology of stomata. *Carnegie Institution of Washington Publication* **82**, 1–142.

Lösch, R. (1977). Responses of stomata to environmental factors I. Temperature and humidity. *Oecologia* **29**, 85–97.

MacDowall, F. D. H. (1963). Midday closure of stomata in ageing tobacco leaves. *Canadian Journal of Botany* **41**, 1289–1300.

Mansfield, T. A. & Heath, O. V. S. (1963). Studies in stomatal behaviour IX. *Journal of Experimental Botany* **14**, 334–52.

Marrè, E., Lado, F., Caldogno, F. & Colombo, R. (1974). Fusicoccin-activated proton exclusion coupled with K^+ uptake. *Academia Nazionale Del Lincei Serie* **57, b**, 690–700.

Maskell, E. J. (1928). Experimental researches in vegetable assimilation, XVIII. *Proceedings of the Royal Society of London*, Series B **102**, 488–533.

Meidner, H. (1955a). Changes in the resistance of the mesophyll tissue with changes in leaf water content. *Journal of Experimental Botany* **6**, 94–99.

Meidner, H. (1955b). The determination of paths of air movement in leaves. *Physiologia Plantarum* **8**, 930–5.

Meidner, H. (1965a). A simple porometer for measuring the resistance to air flow offered by stomata of green leaves. *School Science Revue* **47**, 149–51.

Meidner, H. (1965b). Stomatal control of transpirational water loss. In *19th Symposium of the Society for Experimental Biology*, pp. 185–203. Cambridge University Press.

Meidner, H. & Spanner, D. C. (1959). The differential transpiration porometer. *Journal of Experimental Botany* **10**, 190–205.

Meidner, H. & Bannister, P. (1979). Pressure and solute potentials in stomatal cells of *Tradescantia virginiana*. *Journal of Experimental Botany* **30**, 255–65.

Meidner, H. & Edwards, M. (1975). Direct measurements of turgor pressure potentials, I. *Journal of Experimental Botany* **26**, 319–30.

Milburn, J. A. (1979). An ideal viscous flow porometer. *Journal of Experimental Botany* **30**, 1021–34.

Milthorpe, F. L. (1955). The significance of the measurements made by cobalt paper method. *Journal of Experimental Botany* **6**, 17–19.

Molisch, H. (1912). Das Offen und Geschlossensein der Spaltöffnungen (Infiltrations Methode). *Zeitschrift für Botanik*, 106–7.

Moody, W. & Zeiger, E. (1978). Electrophysiological properties of onion guard cells. *Planta* **139**, 159–65.

Moreshet, S., Stanhill, C. S. & Koller, D. (1968). A radioactive tracer technique for the measurement of diffusion resistance. *Journal of Experimental Botany* **19**, 460–7.

Moreshet, S. & Stanhill, C. S. (1965). The relation between leaf mass flow resistance and infiltration score. *Annals of Botany, N.S.* **29**, 625–33.

Moreshet, S. & Yocum, C. S. (1972). A condensation type porometer for field use. *Plant Physiology* **49**, 944–9.

Mouravieff, I. (1955). Action du CO_2 et de la lumière sur le stomate separé du mesophylle, I. *Comptes Rendus, Académie Sciences, Paris* **242**, 926–7.

Mouravieff, I. (1956). Action du CO_2 et de la lumière sur l'apparail stomatique separé du mesophylle. *Botaniste, ser.* **40**, 195–212.

Müller, N. C., (1869). Untersuchungen über die Diffusion atmosphärischer Gase. *Jahrbücher für Wissenschaftliche Botanik* **7**, 145–92.

Müller, N. C. (1872). Die Anatomie und die Mechanik der Spaltöffnungen. *Jahrbücher für Wissenschaftliche Botanik* **8**, 75–116.

Ogunkanmi, A. B.,Tucker, D. J. & Mansfield, T. A. (1973). An improved bioassay for ABA and other antitranspirants. *Planta* **115**, 47–53.

Outlaw, W. H. & Lowry, O. M. (1977). Organic acid and K^+ accumulation in guard cells during stomatal opening. *Proceedings of the National Academy of Sciences, USA* **74**, 443–4.

Outlaw, W. H. & Manchester, J. (1979). Guard cell starch accumulation quantitatively related to stomatal aperture. *Plant Physiology* **64**,79–82.

Penman, H. L. (1942). Theory of porometers used in the study of stomatal movements in leaves. *Proceedings of the Royal Society of London*, Series B **130**, 416–33.

Penman, H. L. & Schofield, R. K. (1951). Some physical aspects of assimilation and transpiration. In *5th Symposium of the Society for Experimental Biology*, pp. 115–29. Cambridge University Press.

Raschke, K. (1965). Eignung und Konstruktion registrierende Porometer für das Studium der Schliesszellen Physiologie. *Planta* **67**, 225–41.

Raschke, K. & Fellows, M. P. (1971). Stomatal movements in *Zea mays*; shuttle of K^+ and Cl^- between guard cells and subsidiary cells. *Planta* **101**, 296–316.

Sampson, J. (1961). A method of replicating dry or moist surfaces for examination by light microscopy. *Nature* **191**, 932.

Schnabl, H., Bornman, C. H. & Ziegler, H. (1978). Studies on isolated starch containing and starch deficient guard cell protoplasts. *Planta* **142**, 33–9.

Schönherr, J. & Bukovac, M. J. (1972). Penetration of stomata by liquids. *Plant Physiology* **49**, 813–19.

Sheriff, D. W. & McGruddy, E. (1976). Changes in leaf viscous flow resistance following excision measured with a new porometer, *Journal of Experimental Botany* **27**, 1371–5.

Shreve, E. B. (1919). The role of temperature in the determination of the transpiring power of leaves by hygrometric paper. *Plant World* **22**, 172–80.

Slatyer, R. O. & Jarvis, P. G. (1966). A gaseous diffusion porometer for continuous measurement of diffusive resistance of leaves. *Science* **151**, 574–6.

Spanner, D. C. (1953). On a new method of measuring the stomatal aperture of leaves. *Journal of Experimental Botany* **4**, 283–95.

Spanner, D. C. & Heath, O. V. S. (1951). Some sources of error in the use of the resistance porometer and some modifications of its design. *Annals of Botany, N.S.* **15**, 319–31.

Squire, G. R. & Mansfield, T. A. (1972). A simple method of isolating stomata on detached epidermis by low pH treatment. *New Phytologist* **71**, 1033–43.

Stahl, E. (1894). Einige Versuche über Assimilation und Transpiration. *Botanische Zeitung* **52**, 117.

Stålfelt, M. G. (1959). The effect of CO_2 on hydroactive closure of stomata, *Physiologia Plantarum* **12**, 691–705.

Stålfelt, M. G. (1963). Diurnal dark reactions in the stomatal movements. *Physiologia Plantarum* **16**, 756–66.

Turner, N. C., Pedersen, F. C. C. & Wright, W. H. (1969). An aspirated diffusion porometer for field use. *Connecticut Agricultural Experiment Station Special Bulletin* **29**, 200.

Wallihan, E. F. (1964). Modification and use of an electric hygrometer for estimating relative stomatal apertures. *Plant Physiology* **39**, 86–90.

Weatherley, P. E. (1966). A porometer for use in the field. *New Phytologist* **65**, 376–87.

Weber, F. (1927). Stomata Öffnungszustand bestimmt mit Cellophan. *Berichte Deutschen botanischen Gesselschaft.* **45**, 534–6.

Williams, W. T. & Barrett, F. A. (1954). The effect of external factors on stomatal starch. *Physiologia Plantarum* **7**, 298–311.

Willmer, C., Pallas, J. E. & Black, C. C. (1973). CO_2-metabolism in leaf epidermal tissue. *Plant Physiology* **52**, 448–52.

Wong, S. C., Cowan, I. R. & Farquhar, G. D. (1979). Stomatal conductance correlates with photosynthetic capacity. *Nature* **282**, 424–26.

Wormer, T. M. & Ochs, R. (1959). Humidité du sol, ouverture des stomates et transpiration du palmier a huile et de l'arachide. *Oléagineux* **14**, 571–80.

Zeiger, E. & Hepler, P. K. (1977). Light and stomatal function: blue light stimulates swelling of guard cell protoplasts. *Science* **196**, 887–9.

E. A. C. MACROBBIE

Ionic relations of stomatal guard cells

Introduction

The importance of ion movements in producing the turgor changes in guard cells responsible for the opening and closing of stomatal pores is now clearly recognized, and the evidence implicating the accumulation of potassium salts in the guard cells in the opening process has been discussed in a number of recent reviews (Raschke, 1975, 1979; Thomas, 1975; Hsiao, 1976; MacRobbie, 1977). Although the end results of the processes leading to stomatal opening have been established in a number of species, in a semi-quantitative form, the nature and sequence of the primary processes involved in the initiation of the ion flux changes leading to these end results are not clearly established, and there is a need for a more detailed quantitative description of the associated changes in the ionic state of the guard cells.

In the intact plant accumulation of potassium salts is generally involved, but the degree to which potassium is balanced by chloride taken into the cell, and by malate or other organic acid anion synthesized within the cytoplasm, depends on the species and the experimental conditions.

The aim of this review is to consider the quantitative information available on the changes in the ionic state of guard cells associated with opening and closing, and the extent to which this throws light on the nature of the primary processes, and on the mechanisms by which the guard cell responds to environmental signals in the control of aperture. An important aim in any discussion of this kind must be to identify ways in which the ionic relations of guard cells are special, and peculiar to guard cells, differing from those of other vacuolate plant cells, including cells in the leaf epidermis.

Our aim must be to provide a quantitative description of the ionic state of guard cells, as a function of aperture. This implies measurements of internal and external ion concentrations, of membrane potentials, of ion fluxes across cell membranes and of intracellular ion synthesis; it also implies separate consideration of the vacuole and the cytoplasm and measurement of tonoplast fluxes as well as those at the plasmalemma. Such information, available both for steady apertures over the range of opening, and for intermediate changing apertures, would allow us to identify the nature of the active processes involved, and the sites of control by environmental signals. At the moment very little of this information is available, and it is difficult to see how some of it could be obtained for stomata in intact leaves. The microenvironmental conditions of guard cells in epidermal strips, bathed with solution in place of the air spaces of the sub-stomatal cavities, may indeed be very different from the conditions in intact leaves, but such stomata still function and offer possibilities of making the measurements required.

Accumulation of salts in guard cells

Although potassium accumulation in open guard cells has been demonstrated histochemically in a very wide range of different species, with a good correlation between aperture and density of precipitate after treatment with sodium cobaltinitrite, we have quantitative information for only a few. Recent work has concentrated on species from which epidermal strips may easily be obtained, largely on *Vicia faba* L., and *Commelina communis* L. Early work on *Zea mays* L. (Raschke & Fellows, 1971) established a shuttle of potassium and chloride between the guard cells and subsidiary cells in this species, with movements of about 0.6 pmol K^+ per complex, of which Cl^- balanced about 40% (but up to 100% in some individual complexes), giving concentrations of about 400 mM K^+ and 150 mM Cl^- in open guard cells. However the greater difficulty of obtaining viable epidermal strips from maize seems to have been judged to outweigh the advantages of a self-contained shuttle between the guard cell and one subsidiary cell, and further work has concentrated on other species.

Vicia faba

Two sets of measurements provide figures for the differences of ion content in guard cells taken from intact leaves with stomata opened in light or closed in darkness. Humble & Raschke (1971), using electron microprobe analysis for potassium, found that an increase of 2.0 pmol K^+ per guard cell associated with opening from 2 to 12 μm; thus the potassium content increased from 0.1 to 2.1 pmol per cell with opening. Using their estimates of single guard cell volumes in this tissue, of 1.3 pl closed and 2.4 pl open, these figures correspond to concentrations of 77 mM closed and 880 mM open. Only about 5% of this change in potassium was balanced by chloride.

By very different methods, microanalysis of single guard cell pairs dissected from frozen dried tissue, Outlaw and co-workers obtained very similar figures (Outlaw & Lowry, 1977; Outlaw & Kennedy, 1978; Outlaw & Manchester, 1979). The changes per guard cell associated with opening to 10 μm were increases of 1.6 pmol K^+, 0.23 pmol malate, 0.145 pmol citrate, 0.42–0.74 pmol hexose and 0.14 pmol sucrose, together with a decrease in starch of 0.25 pmol (hexose equivalents). Thus the organic anions balanced about 62% of the potassium changes. Their figures correspond to potassium concentrations of about 80 mM in closed guard cells and 460–760 mM in open guard cells (using 4 pl as the average guard cell volume). An important feature of these results is the import of material to the guard cells on opening; thus the disappearance of starch accounts quantitatively for the synthesis of organic acid anions, but there are, additionally, increases in both hexoses and sucrose. Opening is associated with solute accumulation, not all of which is potassium salt. The increase in dry weight of guard cells with opening (of about 0.7 ng per guard cell pair) is also consistent with import of solute from the surrounding tissue. It is also worth noting that Outlaw & Kennedy (1978) suggest that there is little malate synthesized in the initial stages of stomatal opening. A more detailed look at guard cell contents at intermediate apertures seems to be needed.

A number of authors have measured changes in the amount of potassium in guard cells during stomatal opening in isolated epidermal strips of *V. faba*, in which the epidermal cells have been killed in or after the stripping process. These results are collected in Table 1. The relation between potassium content and aperture is linear, with slopes, by a variety of methods, of 10 to 33 pmol mm^{-2} per μm of aperture, or 0.1 to 0.25 pmol per guard cell per μm. The figures are similar to those in intact leaves. But one feature of the results, by whatever method, is very striking. Stomata in such strips are partly-open after treatment, and show a linear relation between potassium content and aperture from the partly open to the fully open condition with an intercept on the aperture axis at about 4–5 μm. The implication seems to be that accumulation of potassium salts is responsible for wide opening, but may be less important in the early stages of opening. Again it seems that study of intermediate stages of opening is necessary.

It is clear from these results that in *V. faba* most of the potassium taken up is balanced by the synthesis of organic acid anion within the guard cell, even in epidermal strips opened by floating on potassium chloride solutions; thus, in strips opening on 10 mM KCl + 0.1 mM $CaCl_2$, chloride uptake balanced only 25% of the potassium uptake (Pallaghy & Fischer, 1974), and even on 100 mM KCl, Cl accounted for only 44% of the K uptake (Raschke & Schnabl, 1978). Proton extrusion associated with stomatal opening might therefore be expected, and was measured by Raschke & Humble (1973). They found an output of 0.1–0.55 pmol per cell per μm of aperture, as the excess proton extrusion into potassium solutions, in which stomata opened, over and above that into calcium solutions, in which they did not. The figures are therefore consistent with a 1:1 exchange of K^+ and H^+. The initial release of H^+ into calcium solutions, or from frozen thawed tissue, was very high; it was assumed that this represented free space exchange, and was about 2.7 pmol H^+ per guard cell. This implies a large exchange capacity in guard cell walls, but also an initially low pH in the wall.

Thus it has been established that in guard cells opening in epidermal strips potassium uptake is very largely balanced by malate, and is associated with net proton extrusion; this is true even in the presence of chloride in the bathing solution. What is less clear is whether the same very large proton extrusion takes place in the intact leaf. In the intact leaf such proton extrusion would produce very large changes in pH in the cell walls; for a 10 μm

Table 1. *Changes in ion content and aperture in epidermal strips of Vicia faba*

Ion	Reference	Slope pmol mm$^{-2}\mu$m^{-1}	pmol cell$^{-1}\mu$m^{-1}	Intercept μm
K[a]	Fischer (1972)	26 ± 1 (16)	0.21 ± 0.01	5.2 ± 0.5 (16)
K[a]	Pallaghy & Fischer (1974)	28	0.22	4.5
Cl		7	0.05	5.3
K[b]	Raschke & Schnabl (1978)	23	0.22	5.3
malate		10.3	0.10	–
Cl		1.3	0.01	–
K[c]	Van Kirk & Raschke	10	0.10	3.8
malate	(1978b)	2.5	0.025	–
Cl		4.4	0.044	–
malate[d]	Van Kirk & Raschke	28	0.23	1.7
malate[e]	(1978a, b)	14	0.12	–
K[f]	Shimada et al. (1979)	13	0.13	not determined
malate		5		
K[g]	Allaway & Hsiao (1973)	32.5	0.25	5
K[h]	Allaway (1973)	30.6	0.25	5
malate		9	0.07	6.7

[a] Tracer uptake.
[b] Microprobe analysis for K and Cl; opening on potassium iminodiacetate.
[c] Microprobe analysis for K and Cl; opening on KCl.
[d] Opening on K-iminodiacetate.
[e] Closing on abscisic acid.
[f] Isotachophoretic analysis for K and malate; opened on potassium phosphate buffer, no intermediate apertures.
[g] Flame photometry; opening/closing by treatment of leaf before stripping.
[h] Opening/closing by treatment of leaf before stripping.

opening, an output of 4 pmol H^+ per guard cell pair into the limited cell wall space would produce a very low pH, and diffusion of protons away from the site of production is likely to be limiting. In the intact leaf it may well be that the falling pH outside the neighbouring cells leads to solute leakage from these cells. The guard cells and neighbouring cells must be in direct competition for solutes in their common free space (which is of limited volume). If the guard cells are less sensitive to low pH than are neighbouring cells (for which we have evidence in *Commelina communis*) then proton extrusion by guard cells may tilt the competitive balance in their favour. Movement of malate from neighbouring cells, rather than synthesis in the guard cell, may play a larger role in the intact leaf than in epidermal strips bathed by an infinite volume of bathing solution.

Commelina spp.

Various estimates of potassium content of the epidermis removed from leaves treated in light or dark to open or close the stomata agree in the conclusion that there is no change in the total amount of potassium with changes in aperture. In *C. cyanea* R.Br., Pearson (1975) found 3500 ± 17 pmol mm^{-2} for the epidermal potassium content, not correlated with aperture. In *C. communis* Bowling (1976) found no significant differences in the concentrations of potassium, chloride and malate in extracts of epidermis taken from leaves in light or dark, with concentrations of 140–147 mM K, 48–49 mM Cl and 61–64 mM malate. These give very high values for the total ion content; if the epidermis has a volume of 30–40 nl nm^{-2} then the contents are about 4300–5800 pmol K^+ mm^{-2}, 1500–1900 pmol Cl^- mm^{-2} and 1900–2500 pmol malate mm^{-2}. In later measurements potassium contents were determined by flame photometry on extracts of epidermal strips removed from leaves of *C. communis* treated in light or dark, with open and closed stomata respectively; the values were 2720 ± 160 (14 replicates) pmol mm^{-2} in light, and 2620 ± 250 (16) after dark treatment, and the ratio of contents in paired strips from sections of the same leaf kept in light and dark was 1.08 ± 0.11 (7). (MacRobbie, unpublished observations).

Amounts of potassium and chloride, and pH, have been estimated in individual cells of the epidermis by the insertion of ion-sensitive microelectrodes to measure ion activity (Penny & Bowling, 1974, 1975; Penny, Kelday & Bowling, 1976; See Table 2). The results suggested a shuttle of potassium and chloride between cells of the epidermis, into the guard cells on opening, and out of guard cells into other epidermal cells on closing. Similar figures for potassium have been obtained in subsequent work, with a different type of potassium electrode, discussed in a later section of this paper (MacRobbie & Lettau, 1980, and unpublished observations). An important point about the results in Table 2 was made by Allaway (1976), who pointed out that much more potassium is lost from the very large epidermal cells on stomatal opening than appears in the much smaller cells of the stomatal complex. Thus the figures would require extracellular storage of potassium, in amounts dependent on aperture. In this respect the stomatal sacs in *C. communis* reported by Stevens & Martin (1977) might be important as potassium storage sites, but this would imply that considerable extracellular ion exchange (of K^+ for Ca^{2+}?) is involved in opening.

The figures in Table 2 show that although chloride plays a part in the salt accumulation in open guard cells of *C. communis*, a considerable fraction of the potassium must be balanced by organic acid anions. A number of authors have measured malate in the epidermis, and these figures are collected in Table 3. They are all much lower than the figure already quoted, that calculated from Bowling's extract (1976). The figures for epidermal strips of *C. communis* taken from intact leaves, particularly those of Van Kirk & Raschke (1978a, b), suggest that cells other than guard cells must contribute a significant fraction of the epidermal malate, which is high even in the dark, when stomata are closed. This would be expected from the estimates of

Table 2. *Ion concentrations[a] in various cells of the epidermis*

Cell[b]	Open			Closed		
	K	Cl	pH	K	Cl	pH
Commelina communis (Penny & Bowling, 1974, 1975; Penny, Kelday & Bowling, 1976)						
GC	448	121	5.60	95	33	5.19
ILS	293	62		156	36	5.60
OLS	98	47	5.56	199	55	5.78
TS	169			289		
EC	73	86	5.11	448	117	5.74
Tradescantia albiflora (Zlotnikova, Gunar & Panichkin, 1977)						
GC	633			152		
LS	73			194		
TS	106			116		
EC	58			56		

[a] Concentrations (in mM) estimated from calibration of K-electrode and Cl-electrode in KCl solutions.

[b] GC, guard cell; ILS, inner lateral subsidiary cell; OLS, outer lateral subsidiary cell; TS, terminal subsidiary cell; EC, epidermal cell.

potassium and chloride in Table 2, which show that in all cells a considerable amount of potassium must be balanced by organic acid anions rather than by chloride. In strips on solution the malate contents are strongly correlated with aperture as stomata open, and are much lower than those in intact leaves. It seems likely that malate is lost from epidermal cells in strips floating on solution, and that malate synthesis in the guard cell may play a much more important role in opening of stomata in floating strips than it does in the intact leaf – as was suggested also in the discussion of the results for *V. faba*.

Bowling (1976) suggested that a shuttle of potassium ions and malate between guard cells and other cells of the epidermis was involved in stomatal opening and closing. The existence of such a shuttle seems likely, although the mechanism he proposed for its initiation seems less so. He suggested that a pre-closing fall in the guard cell pH would convert a large fraction of divalent malate to monovalent malate, which could then diffuse out of the guard cells to the other cells of the epidermis; he argued that divalent malate was locked into cells by its inability to cross membranes. He postulated that a pre-opening rise in guard cell pH would convert monovalent malate to

Table 3. *Changes in epidermal malate with stomatal aperture in Commelina spp.*

Species	Reference	Slope pmol mm^{-2} μm^{-1}	Malate content open (15 μm)	pmol mm^{-2} closed
Opening/closing on intact leaves in light/dark				
C. communis	Dittrich *et al.* (1979)	10–25	550–600	100–200
C. communis[a]	Van Kirk & Raschke (1978a)	5, 10	551, 381	474, 243
C. cyanea	Pearson (1973)	4.3	90	25
C. cyanea	Pearson & Milthorpe (1974)	3.7	116	60
C. cyanea	Pearson (1975)	11.4	200	30
Opening/closing in floating epidermal strips				
C. communis[b]	Van Kirk & Raschke (1978a)	25, 19	551, 381	162, 117
C. communis[c]	Travis & Mansfield (1977)	12.5	180	(20)
Guard cell malate which would balance excess K in Table 2				
C. communis			156	18

[a] Leaf in light or dark to open/close.
[b] Stomata initially open, when removed from leaf, closing on ABA.
[c] Stomata opening, floating on buffer.

divalent malate, locking it into the guard cell, and setting up a concentration gradient of monovalent malate for its diffusion from epidermal cells into the guard cell. While such pH changes, if they preceded any other changes, would induce large shifts in the balance of malate$^-$: malate^{2-} (though not as large as Bowling suggests, since we must conserve moles of malate, and not equivalents, in the initial stages), it seems unlikely that diffusion down gradients between vacuoles is responsible for any shuttle, for two reasons. The largest changes in pH take place, inversely, in the guard cell vacuoles and in the epidermal cell vacuoles. These are connected through the respective cytoplasms and the extracellular space. It seems unlikely that the cytoplasmic pH of either cell is subject to such drastic changes of pH, but is more likely to be regulated near neutrality. We would not therefore expect large shifts in malate movements into and out of cells, although we might expect large changes at the tonoplasts. Secondly we must discuss ion fluxes in terms of gradients at the plasmalemmas and tonoplasts, considered separately, and not simply in terms of the gradients from cell to cell. Transfer between rooms on the upper floors of two separate buildings requires either effort or a working lift in the second, even if the final levels are the same. Raschke (1977) has suggested that increasing malate levels in the cytoplasm (and/or low pH) might induce leakage of K^+, Cl^- and malate from guard cells. If the malate–switch hypothesis has a role in stomatal mechanisms it would seem likely that it concerns malate movements from vacuole to cytoplasm initially, rather than at the plasmalemma, and might have a role in determining the level of cytoplasmic malate. However the existence of guard cells operating on potassium chloride accumulation rather than potassium malate accumulation, such as maize (Raschke & Fellows, 1971) and onion (Schnabl & Ziegler, 1977), make it unlikely that malate transfer out of the vacuole is an essential feature of the steps leading to closing.

It is unknown whether the large change in vacuolar pH precedes cytoplasmic changes and then causes them, or whether it is one of the results of changes in cytoplasmic processes, such as proton pumping at the plasmalemma, for example. It seems important to establish the timing of the various changes in the ionic state of the guard cell associating with opening, and to discover which comes first – changes in cytoplasmic pH and plasmalemma potential, changes in ion fluxes at the plasmalemma and/or tonoplast, changes in vacuolar pH.

Finally it is worth noting that although Dittrich, Mayer & Meusel (1979) found concurrent cycling of epidermal malate and stomatal aperture in epidermis on intact leaves in light/dark cycles, the time courses differed; malate content continued to increase after stomata had opened to their maximum aperture, and continued to fall after the stomata had closed.

Again it would seem worthwhile to look at intermediate apertures, and at transients, in more detail than has yet been done.

Tradescantia albiflora Kunth var. *albo-vittata*

Measurements of potassium content have also been made, in guard cells, subsidiary cells, and epidermal cells of *T. albiflora*, in detached leaves or in intact plants, with open and closed stomata (Zlotnikova, Gunar & Panichkin, 1977*a*). The figures have been used to construct a balance sheet for potassium movements, and it appears that in this species the increase in guard cell potassium on opening can be accounted for by the decrease in the lateral subsidiary cell; there was little change in the potassium levels in terminal subsidiary cells or in epidermal cells. The figures for guard cells were 633 ± 50 (7 replicates) mM with open stomata, and 152 ± 8 (12) mM when stomata were closed (full figures in Table 2).

Conclusions

The results already discussed show that guard cells are capable of considerable vacuolar salt accumulation during stomatal opening. Most plant cells regulate their salt accumulation, often in response to a pressure signal as turgor increases, and shut down net salt uptake at much lower levels of accumulation than we observe in guard cells of open stomata (Cram, 1976; Zimmermann, 1977). But although guard cells seem capable of accumulation to much higher levels than other cells, there is little indication that the processes involved are special, or in any way different from those of other cells. Salt accumulation is commonly held to arise from primary proton extrusion, generating an electrical force for potassium entry, and a pH gradient at the plasmalemma; the increasing alkalinity in the cytoplasm may then be reversed by synthesis of malic acid in the cytoplasm or by chloride uptake by co-transport with hydrogen ions. It is essential to recognize that for either of these processes to continue the removal of salt (either potassium chloride or potassium malate) to the vacuole is essential; chloride entry is feedback-inhibited by internal chloride in many cells (reviews already cited; Cram, 1973; Sanders, 1980), and phospho-enol pyruvate carboxylase is feedback-inhibited by malate (Ting, 1968). It may be difficult to measure tonoplast transport processes, but their role in overall control should not be underestimated. The processes demonstrated in guard cells are entirely consistent with this general pattern of salt accumulation, but it is clear that our information is quite inadequate to establish the nature of the process involved. To do so, we require measurements of the driving forces and fluxes. Thus we require measurements not only of vacuolar concentrations, but also of cytoplasmic and extracellular concentrations, and of membrane potentials at both plasmalemma and tonoplast. We have some measurements of membrane potential either in intact leaves or in epidermal strips bathed in defined solutions, which will be discussed in the next section, but we have no information on the extracellular concentrations in the guard cell walls in intact leaves. It is difficult to see how this could be measured, and it seems, therefore, that the identification of the nature of the transport processes, their characterization as primary active transport, secondary active transport or passive transport, will depend upon measurements on epidermal strips in defined bathing media. Their condition may differ from those in the intact leaf, but it seems the only prospect of a well-defined system for study. It may then be possible, for example, to decide whether potassium is ever accumulated against its electrochemical potential gradient, by primary active transport, or whether an electrical gradient set up by electrogenic proton pumping can always provide an adequate driving force for potassium entry.

Membrane potentials in guard cells

There are very few measurements of membrane potentials in guard cells, and none that can, with any certainty, be combined with measurements of internal and external concentrations of potassium, to determine the direction of the electrochemical potential gradient for potassium at the plasmalemma. Pallaghy (1968) measured membrane potentials in guard cells in epidermal strips of tobacco; in 1 mM KCl outside the potential was -110 ± 3 (6 replicates) mV, and in (1 mM NaCl + 10 mM KCl) the value was -76 ± 1.5 (20) mV. Thomas (1970) found apertures of about $5.5 \mu m$ in 1 mM KCl as bathing solutions, but we have no values for the internal potassium concentration under comparable conditions. Sawney & Zelitch (1969) found a linear relation between aperture and potassium content in the guard cells of tobacco, by electron microprobe analysis of epidermal strips removed from intact leaves; the internal concentration at $5.5 \mu m$ was about 400 mM. The potassium concentration which could be achieved passively with 1 mM outside and 110 mV potential difference is only 79 mM. However, because stomata in epidermal strips on solution are likely to open wider for a given content than those in intact leaves, we cannot be certain that active transport of potassium is required; the results make it likely, but the argument is shaky.

Membrane potential measurements in guard cells of *Tradescantia albiflora* were made by Gunar, Zlotnikova & Panichkin (1975) and Zlotnikova, Panichkin & Gunar (1977), both in intact plants and in detached leaves. With stomata closed the membrane potential was about -170 mV, whether measured on intact plants (with the reference electrode in the root bathing solution), or in detached leaves (with the reference electrode in the solution bathing the cut end). In intact plants, with open stomata, a potential of − 87 mV was recorded. One problem with these measurements is the likelihood of differences in potential between the root bathing solution and the leaf apoplast, particularly in transpiring plants, and therefore the figure for open stomata may not be a good measure of the transmembrane potential at the guard cell. There is also uncertainty in the appropriate value for the external concentration; a concentration of 0.5 to 2.5 mM K in the bathing solution seems to have been used, but the concentration in the guard cell wall may be very different. Zlotnikova *et al* (1977*b*) also investigated the effect of light/dark transitions on the guard cell membrane potential, recording a depolarization of 9–17 mV on darkening, and hyperpolarization on subsequent re-illumination.

Membrane potentials have also been measured in onion guard cells (Zeiger, Moody, Hepler & Varela, 1977; Moody & Zeiger, 1978). With 1 mM K outside, in dark-adapted guard cells, impaled in dim green light, they found a membrane potential of − 72 mV; in short light/dark transitions they observed depolarization of an average 23 mV in the dark, and hyperpolarization of an average 10 mV in light. They suggest that photo-activation of an electrogenic proton pump is the most likely cause of the potential changes observed. Again this is not a special characteristic of guard cells, but is typical of ordinary green cells, in a range of plants.

Saftner & Raschke (quoted by Raschke, 1979) have measured membrane potentials in guard cells of a number of species, with values of − 42 to − 50 mV when the bathing solution contained 30 mM KCl, which were responsive to the external concentration of alkali salts in the range 3–30 mM. They also observed a cell wall potential, typical of a Donnan space with a fixed negative charge density of 0.3–0.5 M.

Measurements of the membrane potentials in the cells in intact epidermal strips of *Commelina communis*, bathed in 0.1 mM KCl, 0.1 mM NaCl, 1 μM $CaCl_2$, gave figures of − 89 ± 7 (11 replicates) mV for closed guard cells, and − 99 ± 11 (4) mV for open guard cells (MacRobbie & Lettau, unpublished); in one strip in which sodium chloride was added to this bathing solution the guard cell membrane potential depolarized by 22 mV per decade between 0.1 and 10 mM, and 65 mV between 30 and 67 mM. Again these results are not strikingly different from the behaviour of other plant cells.

Recent studies on *Commelina communis*

The results discussed so far suggest a number of pressing questions, to which answers are required before we can hope to understand the nature of the processes involved in stomatal opening and closing. Four main gaps in our knowledge may be considered as a start.

(1) We need to have measurements of internal ion concentrations, under conditions of steady aperture in which we can measure potentials at defined external concentrations, i.e. in epidermal strips. We can then establish the directions of the gradients of electrochemical potential at the cell membranes. Figures for the electrochemical potential gradients between different cell vacuoles are not an adequate basis for discussion.

(2) We need to have measurements of all kinds, over the whole range of aperture, and not simply in open and closed conditions. We need to establish whether the processes involved in the first stages of opening are the same as those responsible for wide opening.

(3) We need flux measurements under defined conditions, at various steady apertures, but we also need to establish which fluxes change as the primary response to opening or closing signals. We need measurement of fluxes at both plasmalemma and tonoplast.

(4) We need to establish the time sequence of the various changes which take place – changes of potential, in potassium concentrations, in pH (outside, cytoplasmic and vacuolar), and of ion fluxes in both directions at both membranes.

When we consider the feasibility of making such measurements it is clear that epidermal strips are necessary, in which we have control of the external environment of the guard cell, and that for flux measurements the strips should contain guard cells as the only live cells, with no intact epidermal/subsidiary cells. The rest of this paper discusses recent work on *C. communis*, done with these long-term aims, but representing only a progress report on the start of such a programme. *C. communis* was chosen because of the large amount of previous work on this species, and because it is possible to produce 'isolated' guard cells, in epidermal strips in which all cells other than guard cells have been killed by treatment at low pH (Squire & Mansfield, 1972).

The experimental work discussed is of two kinds. The first concerns measurements of potassium concentrations in guard cells over a range of apertures, in both 'intact' epidermal strips (with all cells still alive) and in 'isolated' epidermal strips; the osmotic effects of the changes in potassium concentration are compared with the osmotic effects required to change the aperture. The conclusion is that the measured potassium changes are not always adequate to account for the changes in aperture. The second part looks at the effects of abscisic acid (ABA) on ion fluxes in guard cells, and suggests that an increase in ion efflux, rather than an inhibition of ion influx, may be responsible for the ABA-induced closure of stomata.

Potassium changes and osmotic changes in guard cells

Potassium activity measurements were made in the various cells of the epidermis, by insertion of a double-barrelled potassium-sensitive electrode, with Corning liquid ion-exchanger 477317 in one barrel. The electrodes were calibrated before and after insertion, using solutions of potassium chloride. The measurements were made quickly after removal of strips from leaves treated in light or dark to vary the aperture (MacRobbie & Lettau, unpublished).

Similar measurements were made in 'isolated' guard cells, in epidermal strips treated at low pH, to kill all cells other than the guard cells. Stomata in such strips open when floated on KCl (or RbCl or KBr), to about 4–8 μm in 10 mM, about 8 μm in 30 mM, 11–15 μm in 60 mM, and 13–16 μm in 90 mM; there is little difference between pH 3.9 and pH 6.7, at least for exposure times up to 45 h. Ion concentration changes in such 'isolated' guard cells can also be estimated by tracer uptake experiments, by measuring the amount of intracellular ^{86}Rb or ^{82}Br after loading for long enough to reach steady aperture, and full exchange of internal ion content with external tracer. The extracellular activity is first removed in an initial rinse, for a time determined from the efflux curve. The free space exchange for ^{86}Rb was slow at pH 6.7, and therefore the experiments were done at pH 3.9. This pH was also used for the potassium-electrode experiments in 'isolated' guard cells. By combining tracer contents thus measured (in pmol mm^{-2}) with estimates of stomatal frequency and guard cell volumes, another estimate of the concentrations may be obtained, and compared with the results of the electrode measurements (MacRobbie & Lettau, 1980).

The osmotic changes required to change the aperture may be estimated from observations of the decreases in aperture as sucrose is added to the solution bathing epidermal strips; all solutions contained 50–75 mM KCl to avoid solute leakage from the guard cells as the external osmotic pressure was increased (MacRobbie, 1980). For such changes we have the relation:

$$- d\pi_o = d\psi_c = dP - dQ/V + (Q/V^2) \cdot dV$$

where π_o is the external osmotic pressure, ψ_o and ψ_c the external and internal water potentials, P the cell pressure, V the cell volume, and Q the cell solute content. Under experimental conditions Q is taken to be constant, and $-d\pi_o$ is then equal to $dP + (Q/V^2) \cdot dV$; but this quantity will also be equal to dQ/V for changes at constant water potential. Therefore we may use the slope

$(d\pi_o/dA)$ under the experimental conditions as an estimate of $(1/V)$ (dQ/dA), the change in internal solute content needed to open or close the pore at constant water potential. We may then compare such estimates with the measured potassium changes, and assess the extent to which potassium salt accumulation can account for the osmotic changes required.

Potassium in 'intact' guard cells

Potassium activity (K_{ac}) measurements were made over a range of apertures. In the guard cells K_{ac} increased from 50–60 mM in closed stomata, to 260–390 mM at an aperture of 14 μm, but most of the increase took place over the later stages of opening, above 10 μm. Smooth curves were drawn for the relation between K_{ac} and aperture, and further calculations used values read from such curves. Two sets of measurements were made, on plants grown under rather different conditions, and both sets of figures are quoted in Table 4. In the calculations K_{ac} was converted to concentration (K_c) by using the activity coefficient in KCl, and the osmotic potential of this concentration of KCl was then calculated. This is an uncertain conversion, since a fraction of potassium is balanced by malate rather than chloride. To some extent the errors involved in this approximation cancel each other; calibration of the electrode in solutions containing both malate and chloride show that K_c will be underestimated in the presence of malate if the potassium chloride calibration is used, but the lower osmotic value of the malate

Table 4. *Potassium content and aperture in* Commelina communis

	Aperture (μm)			
	2	6	10	14
INTACT strips				
K_{ac} mM[a]	60–71	84–102	146–168	256–390
K_c mM[b]	75–92	110–138	204–237	385–615
π_{KCl} mOsmol kg^{-1}	135–162	200–255	375–440	690–1110
ISOLATED strips				
K_{ac} mM		44	58	98
K_c mM		54	73	130
$^{86}Rb_c$ mM		42	110	162
π_{KCl} mOsmol kg^{-1c}		100–75	130–200	245–295

[a] Two sets of measurements, numbers read off smooth curves for K_{ac}, drawn through points.

[b] Concentration K_c, calculated from activity K_{ac} from calibration in KCl solutions.

[c] The first figure comes from electrode measurements, the second from tracer measurements.

solution will then counteract this. With one-third of potassium balanced by malate the osmotic values will be underestimated by about 25%, but with two-thirds balanced by malate the underestimate will be by about 50–60%. Potassium was estimated in digests of such strips, by the electrode method and by flame photometry. Similar values were obtained by the two methods, and hence it is felt that the electrode measurements are not seriously in error.

Potassium in 'isolated' guard cells

Potassium activity measurements were made in guard cells in 'isolated' epidermal strips floated overnight on different concentrations of KCl (pH 3.9) to vary the aperture. The results are shown in Table 4. In the tracer experiments 'isolated' strips were floated overnight on varying concentrations of ^{86}RbCl, and the total tracer content Q^* (pmol mm^{-2}) was determined by counting strips after a rinse to remove extracellular activity. Q^* was correlated with aperture, with slopes of 6.6 to 14.8 pmol mm^{-2} per μm of aperture in six different regressions. In one experiment with K^{82}Br a linear relation was also found, with a slope of 13.6 pmol mm^{-2} μm^{-1}. Thus, as might be expected, potassium entering guard cells floating on potassium chloride solution is balanced by chloride. In 'isolated' guard cells therefore, the calculations based on KCl should not be in error. For further calculation the relation used was: $Q^* = -55 + 13.4A$. The stomatal frequency was 60 mm^{-2}, and the guard cell volume was estimated from photographs taken with Nomarski optics at varying levels of focus; the volume was estimated to fit the regression $V = 4.0 + 0.2A$, with volume in pl, A (aperture) in μm.

The results show that 'isolated' stomata open widely at much lower potassium contents than those required in intact strips, where it is necessary to overcome the resistance to expansion of both the guard cell and the subsidiary cell turgor.

Osmotic measurements

The results of the osmotic experiments are shown in Table 5, which gives figures for the slopes of the curves obtained by measuring changes in aperture as sucrose was added to the bathing solution (50 or 75 mM KCl, pH 6.7 for intact strips, pH 3.9 or 6.7 for 'isolated' strips) (MacRobbie, 1980). Table 5 also gives figures from $d\pi_{KCl}/dA$, calculated from the results in Table 4. The conclusion is that the potassium changes are too small to account for the changes in osmotic potential in 'isolated' strips, and in the earlier stages of opening in intact strips. Potassium salt accumulation is adequate in the wide opening stages of intact strips, and would appear to determine final aperture of well open stomata. But in the early stages, below about $10\,\mu m$, some other process must also be involved.

Effects of ABA on fluxes

The effects of 2×10^{-5}M ABA on both influxes and effluxes have been studied. Its most important effect seems to be a transient stimulation of the efflux of both ^{86}Rb and ^{82}Br, and the inhibition of influx is relatively small. The influx was measured over short times (0.5–1 h) to avoid significant back flux of tracer during the loading period. With ^{86}Rb the influx in 'isolated' strips transferred immediately into the active solution containing ABA was 80, 71 and 87% of the control influx, in three different experiments; the influx in a short loading period in ABA, after 2–3 h pretreatment in unlabelled solution also containing ABA was 75%, 89% and 79% of the control. With ^{82}Br the influx in ABA was 60% of the control. Thus there is an effect, but not a very big one. By contrast the efflux was increased in the period immediately after adding ABA by factors of 1.9 to 7.6 for ^{86}Rb (8 strips), and 5.0 to 12.3 for ^{82}Br (4 strips). The period of high efflux was transient, and after about an hour the rate constant for exchange had again fallen to near the control value, but during this period a significant fraction of the cell tracer is lost.

Table 5. *Changes in* π_{KCl} *compared with the changes in osmotic potential required for stomatal opening:* Commelina communis

| | Change in π_{KCl} (mOsmol kg^{-1}) | |
	Aperture: 7–9 μm	Aperture: 11–13 μm
Intact		
$d\pi/dA$	91 ± 6 (33)	137 ± 7 (20)
$d\pi_{KCl}/dA^a$	34–58	68–88
	38–50	110–226
Isolated		
$d\pi/dA$	74 ± 6 (27)	121 ± 8 (22)
$d\pi_{KCl}/dA^b$	10	20–40
$d\pi_{RbCl}/dA^c$	24–22	20–16

[a] Two sets of measurements on plants grown under different conditions; in each set the first figure relates to the slope at the lower aperture value of the range, the second to the slope at the upper aperture value.
[b] From electrode measurements.
[c] From tracer measurements.
The figures in brackets are numbers of replicates.

Thus the results suggest that the effects of ABA in producing stomatal closure are the result of stimulated efflux of both anions and cations, rather than simply an inhibition of influx, but it is important to recognize that this is a transient stimulation, and not a continuing process. After this initial period the rate constant falls again, and not simply the flux. Various speculations on the mechanism of ABA induced closure have previously been made. Raschke (1977, 1979) has suggested that ABA inhibits the proton pump at the plasmalemma, leading to inhibition of potassium influx and of malate synthesis, or directly enhances malate synthesis; he has also suggested that high malate or low pH might lead to leakage of K^+, Cl^- and malate from guard cells. The small effect on potassium influx now observed might suggest an effect within the cytoplasm (or at the tonoplast) rather than a primary effect on the proton pump, but further, more detailed, experiments will be needed to sort out the processes involved. They appear to involve the excretion of excess (cation + anion) to a new, lower, steady level of accumulated salt, but the mechanism remains unclear.

Conclusions

These results give us more information on guard cell behaviour, and suggest that further work of this kind will contribute to the provision of answers to the questions raised at the start of this section. They suggest that there is a real need for more detailed study of the intermediate stages in guard cell opening, of the ionic states characteristic of part-open guard cells, and of the transient states between one steady aperture and another. They also point to the importance of measuring two-way fluxes and not simply levels of accumulation, in trying to interpret the effects of factors which change the aperture.

The most striking feature of guard cells remains that seen in closure, when they seem to lose the ability to accumulate salts, and to maintain the large central vacuole characteristic of open guard cells – and typical plant cells. It is for this reason that I believe a study of tonoplast fluxes under a range of conditions may throw light on the essential processes involved.

Acknowledgements

The experimental work described in the second part of this paper was done with support from the SRC, whose help is gratefully acknowledged.

References

Allaway, W. G. (1973). Accumulation of malate in guard cells of *Vicia faba* during stomatal opening. *Planta* **110**, 63–70.

Allaway, W. G. (1976). Influence of stomatal behaviour on long distance transport. In *Transport and Transfer Processes in Plants*, ed. I. F. Wardlow & J. B. Passioura, pp. 295–311. New York: Academic Press.

Allaway, W. G. & Hsiao, T. C. (1973). Preparation of rolled epidermis of *Vicia faba* L. so that stomata are the only visible cells: analysis of guard cell potassium by flame photometry. *Australian Journal of Biological Sciences* **26**, 309–18.

Bowling, D. J. F. (1976). Malate-switch hypothesis to explain the action of stomata. *Nature* **262**, 393–4.

Cram, W. J. (1973). Internal factors regulating nitrate and chloride influx in plant cells. *Journal of Experimental Botany* **24**, 328–41.

Cram, W. J. (1976). Negative feedback regulation of transport in cells. In *Encyclopedia of Plant Physiology*, Vol. IIA, ed. M. G. Pitman & U. Lüttge, pp. 284–316. Berlin: Springer Verlag.

Dittrich, P., Mayer, M. & Meusel, M. (1979). Proton-stimulated opening of stomata in relation to chloride uptake by guard cells. *Planta* **144**, 305–9.

Fischer, R. A. (1972). Aspects of potassium accumulation by stomata of *Vicia faba*. *Australian Journal of Biological Sciences* **25**, 1107–23.

Gunar, I. I., Zlotnikova, I. F. & Panichkin, L. A. (1975). Electrophysiological investigations of cells of the stomatal complex of spiderwort. *Soviet Plant Physiology* **22**, 704–7.

Hsiao, T. C. (1976). Stomatal ion transport. In *Encyclopedia of Plant Physiology*, Vol. IIB, ed. M. G. Pitman & U. Lüttge, pp. 195–221. Berlin: Springer Verlag.

Humble, G. D. & Raschke, K. (1971). Stomatal opening quantitatively related to potassium transport. Evidence from microprobe analysis. *Plant Physiology* **48**, 447–53.

MacRobbie, E. A. C. (1977). Functions of ion transport in plant cells and tissues. In *International Review of Biochemistry, Plant Biochemistry*, II, Vol. 13, ed. D. H. Northcote, pp. 211–247. Baltimore: University Park Press.

MacRobbie, E. A. C. (1980). Osmotic measurements on stomatal cells of *Commelina communis* L. *Journal of Membrane Biology* **53**, 189–98.

MacRobbie, E. A. C. & Lettau, J. (1980). Ion content and aperture in 'isolated' guard cells of *Commelina communis* L. *Journal of Membrane Biology* **53**, 199–205.

Moody, W. & Zeiger, E. (1978). Electrophysiological properties of onion guard cells. *Planta* **139**, 159–65.

Outlaw, W. H. Jr.& Lowry, O. H. (1977). Organic acid and potassium accumulation in guard cells during stomatal opening. *Proceedings of the National Academy of Sciences of the USA* **74**, 4434–8.

Outlaw, W. H. Jr. & Kennedy, J. (1978). Enzymic and substrate basis for the anaplerotic step in guard cells. *Plant Physiology* **62**, 648–52.

Outlaw, W. H. Jr. & Manchester, J. (1979). Guard cell starch concentration quantitatively related to stomatal aperture. *Plant Physiology* **64**, 79–82.

Pallaghy, C. K. (1968). Electrophysiological studies in guard cells of tobacco. *Planta* **80**, 147–53.

Pallaghy, C. K. & Fischer, R. A. (1974). Metabolic aspects of stomatal opening and ion accumulation by guard cells in *Vicia faba. Zeitschrift für Pflanzenphysiologie* **71**, 332–44.

Pearson, C. J. (1973). Daily changes in stomatal aperture and in carbohydrates and malate within epidermis and mesophyll of leaves of *Commelina cyanea* and *Vicia faba. Australian Journal of Biological Sciences* **26**, 1035–44.

Pearson, C. J. (1975). Fluxes of potassium and changes in malate within epidermis of *Commelina cyanea* and their relationship with stomatal aperture. *Australian Journal of Plant Physiology* **2**, 85–9.

Pearson, C. J. & Milthorpe, F. L. (1974). Structure, carbon dioxide fixation and metabolism of stomata. *Australian Journal of Plant Physiology* **1**, 221–36.

Penny, M. G. & Bowling, D. J. F. (1974). A study of potassium gradients in the epidermis of intact leaves of *Commelina communis* L. in relation to stomatal opening. *Planta* **119**, 17–25.

Penny, M. G. & Bowling, D. J. F. (1975). Direct determination of pH in the stomatal complex of *Commelina. Planta* **122**, 209–12.

Penny, M. G., Kelday, L. S. & Bowling, D. J. F. (1976). Active chloride transport in the leaf epidermis of *Commelina communis* in relation to stomatal activity. *Planta* **130**, 291–4.

Raschke, K. (1975). Stomatal action. *Annual Review of Plant Physiology* **26**, 309–40.

Raschke, K. (1977). The stomatal turgor mechanism and its response to CO_2 and abscisic acid: observations and a hypothesis. In *Regulation of Cell Membrane Activities in Plants*, ed. E. Marrè & O. Ciferri, pp. 173–83. Amsterdam: North Holland Publ. Comp.

Raschke, K. (1979). Movements of stomata. In *Encyclopedia of Plant Physiology*, Vol. 7, ed. W. Haupt & M. E. Feinleib, pp. 383–441. Berlin: Springer Verlag.

Raschke, K. & Fellows, M. P. (1971). Stomatal movement in *Zea mays*. Shuttle of potassium between guard cells and subsidiary cells. *Planta* **101**, 296–316.

Raschke, K. & Humble, G. D. (1973), No uptake of anions required by opening stomata of *Vicia faba*: guard cells release hydrogen ions. *Planta* **115**, 47–57.

Raschke, K. & Schnabl, H. (1978). Availability of chloride affects the balance between potassium chloride and potassium malate in guard cells of *Vicia faba* L. *Plant Physiology* **62**, 84–7.

Sanders, D. (1980). The control of Cl⁻ influx in *Chara* by cytoplasmic Cl⁻ concentration. *Journal of Membrane Biology* **52**, 51–60.

Sawney, B. L. & Zelitch, L. (1969). Direct determination of potassium ion accumulation in guard cells in relation to stomatal opening in light. *Plant Physiology* **44**, 1350–4.

Schnabl, H. & Ziegler, H. (1977). The mechanism of stomatal movement in *Allium cepa* L. *Planta* **136**, 37–43.

Shimada, K., Ogawa, T. & Shibata, K. (1979). Isotachophoretic analysis of ions in guard cells of *Vicia faba. Physiologia Plantarum* **47**, 173–6.

Squire, G. R. & Mansfield, T. A. (1972). A simple method of isolating stomata on detached epidermis by low pH treatment: observations of the importance of the subsidiary cells. *New Phytologist* **71**, 1033–43.

Stevens, R. A. & Martin, E. S. (1977). Ion-absorbent substomatal structures in *Tradescantia pallidus. Nature* **268**, 364–5.

Thomas, D. A. (1970). The regulation of stomatal aperture in tobacco leaf epidermal strips. I. The effect of ions. *Australian Journal of Biological Science* **23**, 961–79.

Thomas, D. A. (1975). Stomata. In *Ion Transport in Plant Cells and Tissues*, ed. D. A. Baker & J. L. Hall, pp. 377–412. Amsterdam: Elsevier.

Ting, I. P. (1968). CO_2 metabolism in corn roots. III Inhibition of P-enolpyruvate carboxylase by L-malate. *Plant Physiology* **43**, 1919–24.

Travis, A. J. & Mansfield, T. A. (1977). Studies of malate formation in 'isolated' guard cells. *New Phytology* **78**, 541–6.

Van Kirk, C. A. & Raschke, K. (1978a). Presence of chloride reduces malate production in epidermis during stomatal opening. *Plant Physiology* **61**, 361–4.

Van Kirk, C. A. & Raschke, K. (1978b). Release of malate from epidermal strips during stomatal closure. *Plant Physiology* **61**, 474–5.

Zeiger, E., Moody, W., Hepler, P. & Varela, F. (1977). Light-sensitive membrane potentials in onion guard cells. *Nature* **270**, 270–1.

Zimmermann, U. (1977). Cell turgor pressure regulation and turgor pressure-mediated transport processes. In *Integration of Activity in the higher plant, Symposium of the Society for Experimental Biology* **31**, 117–54.

Zlotnikova, I. F., Gunar, I. I. & Panichkin, L. A. (1977a). Measurement of intracellular potassium activity in *Tradescantia* leaf epidermal cells. *Izv. Timiryazev S-Kh. Akad.* **2**, 10–16.

Zlotnikova, I. F., Gunar, I. I. & Panichkin, L. A. (1977b). Light induced changes of the membrane potential in *Tradescantia* leaf epidermal cells. *Izv. Timiryazev S-Kh. Akad.* **3**, 10–14.

Zlotnikova, I. F., Panichkin, L. A. & Gunar, I. I. (1977). Membrane potentials in *Tradescantia* leaf epidermal cells. *Izv. Timiryazev S-Kh. Akad.* **1**, 20–24.

W. G. ALLAWAY

Anions in stomatal operation

Introduction

It is now possible, I think, at the end of the 'seventies, to write a relatively uncontroversial descriptive account of the role of anions in stomatal operation; a subject which had hardly been touched at the start of the decade. This account cannot, however, be a general one, as the data on which it is based come from only the very few species which are most convenient for the study of stomatal physiology. It has turned out that even these few species do not all behave in the same way in this respect, and yet there has been a regrettable tendency to extrapolate from one to the other when specific data are lacking. I shall try to avoid this, in this article, and I will therefore begin by discussing the problem in the context of one of the favourite species.

Observations on different species

Ionic balance in guard cells of Vicia faba *L.*

Much of the change in vacuolar osmotic pressure which leads to increase in turgor and, in turn, to stomatal opening, can be attributed in *Vicia faba* to an increase in potassium ion concentration (Fischer, 1968). The potassium ions are taken up from elsewhere in the leaf (Allaway & Hsiao, 1973) or, in some experimental situations, from the surrounding solution, and it is therefore necessary to postulate that one or more of the following occurrences take place as well: synthesis or uptake of anions; efflux of an osmotically inactive cation; change in pH of the guard cell vacuole; change in electrical potential of the guard cell vacuole. No measurements of electrical potential or of pH of the guard cell vacuole of *Vicia faba* are available. It was shown (Humble & Raschke, 1971) that during stomatal opening *on the leaf* little uptake of chloride ions took place, and in a similar experimental situation about half of the potassium ions taken up by guard cells could be balanced by malate (Allaway, 1973). Other organic acids found to accumulate in epidermis were glycerate and citrate, although these were not specifically localized in the guard cells (Pallas & Wright, 1973).

Potassium uptake and stomatal opening can occur, if conditions are appropriate, in isolated epidermis floated on solutions (Fischer & Hsiao, 1968), and if chloride ions are not provided there is an efflux of hydrogen ions concomitantly with potassium ion uptake (Raschke & Humble, 1973). Such a proton efflux could be accounted for if organic acids were being synthesized inside the isolated guard cells. Malate accumulation in guard cells on epidermal strips of *Vicia faba* has been shown by Van Kirk & Raschke (1978a) to be quite sufficient to balance the probable potassium uptake if chloride ions are absent, although in the presence of chloride ions much less malate accumulated. Considerable chloride uptake was demonstrated by Pallaghy & Fischer (1974). Raschke & Schnabl (1978) subsequently showed that accumulation of chloride may account for half the ionic balance of potassium in guard cells, and malate for the other half, when chloride ion was given at 100 mM. This effect – the reduction of balancing of potassium by malate to only half when chloride is available – has been seen as explaining the half-balancing by malate found by Allaway (1973) in guard cells on whole leaves. Although in that work the plants were grown in a solution containing only 9 μM chloride, the tissues of such plants contained about 23 μmol chloride per gramme fresh mass (Allaway, 1976), some of which was presumably available to guard cells as well as the 0.1 mM chloride ion of the incubation solution. Chloride does not always depress the level of malate accumulated as stomata open, however. In Fig. 1 are collected data from a number of experiments, in which malate was measured in isolated rolled epidermis which had been treated for three hours in solutions with chloride or with only sulphate as anion. The scatter of points does not suggest a different relationship between malate and aperture in the two anion treatments. Differences from the other workers' experiments are in the method of preparing epidermis, in extraction, and in the pH of the treatment solutions (usually pH 7.0, rather than pH 5.6 in Raschke & Schnabl, 1978). The hypothesis put forward by Dittrich, Mayer & Meusel (1979) from their data might be taken as suggesting that low external hydrogen ion concentration could reduce the tendency for chloride to be taken up. Additionally, in Fig. 1 the concentrations of K$^+$ used are all close to the threshold for maximum light-stimulated K$^+$ uptake (about 2 mM, Humble & Hsiao, 1969) in contrast to the 100 mM used in the study where chloride suppression of malate content was observed. At 100 mM KCl stomata of *Vicia faba* will even open in the dark (Humble & Hsiao, 1969):

perhaps Fig. 1 and the experiments of Raschke & Schnabl (1978) were examining different opening mechanisms.

Although much attention has been focused on the accumulation of malate and, lately, chloride, there has also been interest in the possibility that other anions may be involved. Humble & Raschke (1971) were able to discount the accumulation of any substances rich in phosphorus or sulphur in *Vicia faba* guard cells. As already mentioned, Pallas & Wright (1973) found glycerate and citrate to accumulate in epidermis. However, the most significant work in this connection is that of Outlaw & Lowry (1977) in which enzymatic cycling techniques were used to assay for organic acids in single stomata dissected from freeze-dried leaves. They demonstrated an accumulation of citrate and of malate sufficient to balance about 1 pmol potassium ion per guard cell each, with stomatal opening. Little change was observed in isocitrate, glycerate, glutamate or aspartate, and about one-third of the total potassium ion accumulated was left over, presumably to be accounted for by chloride. It is perhaps worth noting, as an aside at this

Fig. 1. Malate level and mean stomatal aperture in isolated epidermal strips of *Vicia faba* floated for 3 h on various concentrations of potassium chloride and potassium sulphate solutions in light and darkness. Hollow symbols are for chloride treatments, filled symbols for sulphate in the absence of chloride. Symbols differ for K^+ concentration: diamonds, 2 mM K^+ unbuffered; squares, 2 mM K^+, pH 7; circles, 4 mM K^+ unbuffered; triangles, 5 mM K^+, pH 7; crosses, 10 mM K^+ as sulphate, pH 7; +, 'initials', strips from the 'pool' of strips kept in the dark on 0.1 mM $CaSO_4$. All 'pH 7' solutions were bicine-buffered, and all solutions were 0.1 mM with respect to Ca^{2+}.

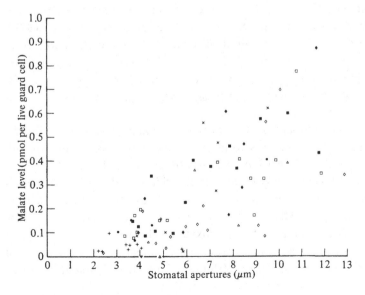

point, that this is another instance of organic anions being insufficient (at least the ones that were measured) to account for all the guard cell ionic balance in the whole-leaf situation. Perhaps because of the microchemical virtuosity required, confirmation of the importance of citrate in stomatal operation has not been forthcoming from other laboratories. Attempts to identify other organic anions in epidermal strips removed from light- or dark-treated leaves has led workers in my own laboratory to conclude that no change in citrate or, as expected, in aspartate, was detectable. In this work the epidermis was rolled and rinsed following removal from the leaf (Allaway & Hsiao, 1973) and while this procedure is likely to remove contaminating contents from other cells of the strip, it is also possible that some ions might be lost from the guard cells (cf. Fischer, 1973). I still feel that further work is required on the identity of ions, apart from potassium and malate, that change during stomatal opening in this species, as well as on the conditions governing the accumulations of Cl^- and other anions.

Source of guard cell malate in Vicia faba

The results already mentioned, showing that guard cells isolated in epidermal strips can accumulate malate, indicate that guard cells can make malate themselves in these circumstances. Malate will only be useful as a balancing ion if it is made from non-ionic precursors, and it has therefore seemed attractive to suggest that the disappearance of starch from guard cells as stomata open (Outlaw & Manchester, 1979) may result from its conversion into organic anions. Further work by the same group has shown that phosphoenolpyruvate carboxylase is present in guard cells in more than sufficient activity to allow for the rate of accumulation of malate required (Outlaw & Kennedy, 1978; Outlaw, Manchester & Di Camelli, 1979). Organic carbon from starch could pass down the glycolytic pathway to phosphoenolpyruvate and then carbon dioxide fixation could lead to malate. Accumulation of label principally in malate has been found where ^{14}C bicarbonate was fed in light or darkness to rolled epidermal strips of *Vicia faba* (Allaway, 1976).

While we may assume, from the above data, that guard cells of *Vicia faba* have the apparatus and ability to make malate themselves, there are still questions to be asked. The first of these is: how are the malate levels of guard cells controlled? Apart from the work on chloride already discussed, there is little direct information in this species. Malate can be formed in carbon dioxide-free air (e.g. Allaway, 1976), which is a little perplexing if it is formed *via* phosphoenolpyruvate carboxylase. The accumulation of malate in guard cells has an action spectrum quite similar to that for stomatal opening and $^{86}Rb^+$ uptake (Ogawa, Ishikawa, Shimada & Shibata, 1978) but there does not seem to be at present a way of connecting this to an hypothesis of control of malate level. There is a need for more detailed time-courses of organic anion accumulation (such as that by Outlaw & Kennedy, 1978) but together with studies of guard cell hydrogen ion release, potassium ion uptake and stomatal opening.

Another question to be asked is: where does the starch come from? Outlaw *et al.* (1979*b*) found no detectable ribulosebisphosphate carboxylase activity in guard cells, and two other Calvin cycle enzymes had very low activities. In the experiment of Allaway (1976) a small amount of label found its way into Calvin cycle intermediates in the light, but it now seems likely that this may have resulted from contamination of the sample with a few photosynthetically-active mesophyll cells, as these are shown up by autoradiography. Malate could be partly converted back to starch on

Fig. 2. Starch content of guard cells of leaflets of *Vicia faba*. Leaflets were kept in the dark overnight, and then either for a further 3 h in darkness in normal air (shown as the first point on the graph), or for various numbers of hours in the light (0.88–1.10 mE m^{-2} s^{-1}, 400–700 nm) in air of reduced carbon dioxide concentration: at the end of the treatment epidermal strips were made, rolled, rinsed, and killed in 80% ethanol. The ethanol was subsequently evaporated off and the starch was hydrolysed and extracted with HCl/DMSO and assayed by measuring the formation of NADH in a coupled assay with amyloglucosidase, hexokinase and glucose-6-phosphate dehydrogenase using a Boehringer starch test kit. Each point shows an individual epidermis sample. Data from Barmby (1979).

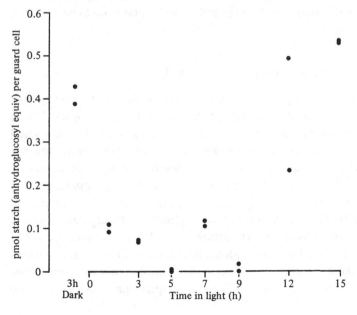

stomatal closure (cf., for example, in crassulacean acid metabolism, Osmond, 1978), or it could be broken down, or it could be allowed to pass out of the guard cells to elsewhere in the leaf. Van Kirk & Raschke (1978*b*) found a loss of malate from epidermal strips of *Vicia faba* in conditions that would close the stomata, but they did not establish conclusively that the malate was released by the guard cells. If malate is lost, and also if it is partly metabolized, the loss of organic carbon to the guard cells could only be made good by import from elsewhere in the leaf, since the guard cells themselves cannot fix carbon dioxide autotrophically. Uptake of labelled sugars from solutions, and conversion into guard cell starch, has been shown in this species (Dittrich & Raschke, 1977*b*).

The negative correlation between guard cell starch level and stomatal opening has been felt by some to be poor: however, if materials for producing starch are imported from elsewhere we should perhaps be able to explain this. Let us suppose that when stomatal opening occurs, guard cell starch is broken down both to provide energy for ion uptake and to be used in the making of malate. At this time there would be little sugar available for translocation to guard cells or anywhere else, so in the first hour or so after the initiation of stomatal opening there should be a good negative correlation between starch in guard cells and stomatal aperture. If observations are continued for longer, however, import of sugar and its reconversion to starch inside the guard cells might begin to take place, even though the guard cells still retain their content of malate and other ions. Late in the day, therefore, there might be no relation between starch content and stomatal aperture. The data for starch in Fig. 2, although rather preliminary, might be said to fit in with this hypothesis.

Anions in stomata of Allium cepa L.

I make no apology for having dealt firstly with my favourite species, nor for mentioning now another on which I have done a little work, before dealing with the several other species that have been studied. *Allium* fits well at this point, too, because the genus does not possess starch at all: this has made it an interesting species to compare with the starchy ones. In *Allium* the reserve carbohydrates are fructans which, although polymers, are soluble in water and do not generally appear as crystals or grains. B. Darbyshire & W. G. Allaway (unpublished) found quite large quantities of 80%-ethanol-soluble fructans up to a degree of polymerization of about 8 or 9 hexose units in dark-treated isolated epidermis of *Allium cepa*, and found that most of these fructans disappeared if the epidermis was given a light treatment. Since the only intact cells in these epidermal strips were guard cells, it is suggested that this finding represents mobilization of soluble oligosaccharide reserves which could be used in stomatal operation. The decrease of fructan was sufficient to account for all the malate found by the same authors to accumulate in chloride-free conditions in the light. It was not shown conclusively, however, that changes in fructan or in malate are involved in stomatal operation in this species. Other authors have found malate accumulation in epidermis of *Allium cepa*, but have not looked for light-induced changes in its concentration (Palevitz & Hepler, 1976; Schnabl, 1978). It is of interest in this context that microelectrode results on guard cells of this species have been taken (although other interpretations are admitted) as favouring an hypothesis of the existence of an electrogenic, outward proton pump which could lead to a passive influx of potassium ions (Zeiger, Moody, Hepler & Varela, 1977; Moody & Zeiger, 1978). Schnabl & Ziegler (1977) found a large increase in both potassium and chloride levels in stomata of epidermis apparently kept in 100 mM KCl in the light, compared with that kept in darkness and without chloride. They suggested that potassium accumulated could be fully balanced by chloride ions, although this has not been demonstrated quantitatively. Frequent treatment of *Allium cepa* plants with chloride during growth was found to inhibit stomatal and guard cell protoplast responses to abscisic acid and to fusicoccin, and to lead to a reduction in the standing malate level of epidermis (Schnabl, 1978).

There are sufficient contradictions in the above data taken together to make it premature to reach a firm conclusion on the anions involved in stomatal operation in *Allium cepa*. Clearly more work is required on this very interesting species.

Anions in stomata of Commelina *species*

Stomatal physiology of four species of *Commelina* has been extensively studied, making this genus perhaps the best-known of all the subjects of stomatal research in recent years. The species studied are quite similar, and I hope it is justified to treat them together. Milthorpe, Thorpe & Willmer (1979) recently published a compendious review dealing principally with these species (although with some data from others) and so it is not necessary for me to do more than summarize.

Uptake of potassium in stomatal opening has been shown in *Commelina* by several authors (e.g. Penny & Bowling, 1974), and the pH of guard cells increases slightly with opening (Penny & Bowling, 1975: *C. communis* L.). Malate accumulation in epidermis has been found with stomatal opening on whole leaves of *C. cyanea* R.Br. (Pearson & Milthorpe, 1974). Guard cells can take up ^{14}C-malate from solutions (*C. communis*: e.g. Dittrich & Raschke, 1977*a*) and this ability has been suggested and discussed as the basis for an ingenious system of operation of stomata in leaves (Bowling, 1976, 1977; Das, Rao & Raghavendra, 1977). However, isolated stomata of the same species can accumulate malate as stomata open, in the absence of any exogenous malate source (Travis & Mansfield, 1977), and ^{14}CO$_2$ fixation principally into malate in isolated guard cells is well known (e.g. Willmer & Dittrich, 1974: *C. diffusa* L.; Willmer & Rutter, 1977; *C. communis*; Thorpe & Milthorpe, 1977, 1979: *C. cyanea*). The time course of malate accumulation and disappearance did not suggest that it was leading the stomatal movement (Dittrich *et al.*, 1979: *C. communis*). Starch content declined during stomatal opening in epidermal strips of *C. benghalensis* L. (Raghavendra, Rao & Das, 1976) suggesting a source for carbon 'skeletons' of organic anions; starch level has been found to increase as stomata of *C. communis* closed (Mansfield & Jones, 1971), and in the same species it has been shown that ^{14}C from exogeneously-fed malate can be rapidly incorporated into starch (Dittrich & Raschke, 1977*a*; Willmer & Rutter, 1977). High levels of phosphoenolpyruvate carboxylase and other necessary enzymes have been demonstrated in *Commelina* epidermis (Willmer, Pallas & Black, 1973: *C. communis*; Thorpe, Brady & Milthorpe, 1978: *C. cyanea*). However, these authors found only little ribulosebisphosphate carboxylase activity, and this suggests that in this species, as in others, guard cells are not autotrophic. There is considerable evidence that in the intact leaf of *Commelina* there is much transfer of organic material from the rest of the leaf to the epidermis, suggesting that isolated epidermal strips are in an abnormal physiological state and therefore that findings with them should be interpreted cautiously (Pearson & Milthorpe, 1974; Dittrich & Raschke, 1977*b*; Willmer, Thorpe, Rutter & Milthorpe, 1978). In *C. communis* (but not in *Vicia faba* or *Tulipa gesneriana* L.) incubation of epidermal strips with glucose in solution has been shown to inhibit stomatal opening (Dittrich & Mayer, 1978), and labelled carbohydrates can be taken up (Dittrich & Raschke, 1977*b*) and incorporated into starch in guard cells. In addition to all these data on malate and related topics, it has been shown that chloride is actively accumulated in guard cells of *C. communis* as stomata open (Penny, Kelday & Bowling, 1976).

Stomatal anions in other species, and some general remarks

Several of the experiments on *Commelina* reported above were repeated by the same workers with *Tulipa gesneriana*. So, for example, we know that tulip epidermis exhibits ^{14}CO$_2$ fixation principally into malate and aspartate (Willmer & Dittrich, 1974; Raschke & Dittrich, 1977). Enhanced activity of phosphoenolpyruvate carboxylase has been found in epidermis of this species (Willmer *et al.*, 1973). As far as I am aware, however, no quantitative measurements of any ion accumulation have been made with tulip. In *Zea mays* L., in contrast, the uptake of potassium (Pallaghy, 1971) has been shown to be accompanied by rather more than one-third as much chloride (Raschke & Fellows, 1971). Organic anions have not been measured in this species, which is rather difficult to work with. There are numbers of related observations on other species (for example, potassium ion uptake and some electrical potentials have been measured in *Nicotiana tabacum* L. (Pallaghy, 1968; Sawhney & Zelitch, 1969)), but there has been insufficient work on those other species to provide results relevant to this discussion.

Trying to sort out the findings on stomatal anions on species lines, as I have done in this paper, has the main advantage that it shows clearly a number of inadequacies in our information. It has, however, the disadvantage of making it very difficult to draw together a model of stomatal ion balance, since it suggests that several models may really be required. It is possible to summarize by pointing out the features that are in common among the various experimental subjects, but unfortunately the common features that are known at present are quite few. I am not sure that the tendency that has been evident in the past to tie together aspects of stomatal operation from one species to another has always been helpful. I should like to suggest that it is time for us to fill in the gaps in the information on the individual species for which we are enthusiasts: we should not necessarily expect to be helped by taking pieces from someone else's jigsaw puzzle and trying to fit them in our own.

Because of the existence of a straightforward histochemical test for potassium ions, potassium accumulation by guard cells is known from a wide range of species (Willmer & Pallas, 1973: Dayanandan & Kaufman, 1973). It seems possible that large numbers of species could be scanned for chloride uptake in a similar way (cf. Schnabl & Ziegler, 1977). The testing for organic anions requires either plant material in which epidermis with 'isolated' stomata, free of contamination with other intact cells, can be prepared, or else micro-dissection and microanalysis techniques. It is possible now to envisage the application of the latter to enable us to survey more widely among species the question of accumulation of organic anions. Both this survey to assess the generality of our knowledge, and the detailed investigations to fill in specific gaps, seem necessary. Before concluding this paper, I will refer to a question related to ion balance in guard cells, but which has scarcely been considered at all.

Ionic content of guard cell cytoplasm and organelles

Cytoplasmic ion content is extremely difficult to measure, since samples of cytoplasm are hard to get: as usual these difficulties are enhanced when dealing with guard cells, which make up only a very small part of any sample of plant material. There are no measurements of the cytoplasmic contents of guard cells, but in other non-halophytic plant cells the major cytoplasmic cation concentration appears to be between 41 and 193 mM (K^+ plus Na^+, with K^+ dominant: from reviews of Hope & Walker, 1975: Wyn Jones *et al*., 1977; Kirst, 1977; and data from Pitman & Saddler, 1967; Jeschke & Stelter, 1976). High ionic strength in the cytoplasm might be expected to interfere with the normal running of the cell, for example by inhibition of enzyme activity (e.g. Greenway & Sims, 1974); such interference is found even in animal cells, where cytoplasmic ionic strengths may be somewhat higher (e.g. Prosser, 1973). I suppose we should think it likely that high inorganic ion content of cytoplasm would inhibit the cells' activities in guard cells, as elsewhere; inhibition of guard cell phosphoenolpyruvate carboxylase by 200 mM NaCl has been shown (Outlaw *et al*., 1979*a*). Yet when the osmotic potential of the vacuole changes during stomatal opening, that of the cytoplasm must change too, since the 'semi-permeable' tonoplast is all that separates them and this has little mechanical strength. Guard cells of closed stomata apparently contain roughly 100–200 mM of inorganic cations, mostly potassium ions (cf. Allaway & Milthorpe, 1975), and so we should expect guard-cell cytoplasm referred to above to be approximately in osmotic balance with this vacuole. When the stoma opens, and the vacuolar ion concentration increases so much, the cytoplasm could (1) dehydrate, so that its osmotic concentration continues to match that of the vacuole, and its volume decreases, (2) increase its ion content by retaining some ions that are passing through to the vacuole, thus retaining its normal volume but increasing its ionic strength, or (3) match its osmotic concentration to that of the vacuole by increasing the concentration of something other than inorganic ions. Humbert & Guyot (1972) and Humbert, Louguet & Guyot (1975) found the ultrastructural appearance of cytoplasm of *Anemia rotundifolia* Schrad and *Pelargonium* × *hortorum* (following fixation, dehydration and staining) to differ between open and closed stomata, and other nonvacuolar changes have been observed in *Vicia faba* (e.g. Heller & Resch, 1967); but it is uncertain whether these changes reflect volume changes *in vivo*. It is possible that guard cells have special cytoplasm capable of tolerating high ion contents, like certain marine algae such as *Valonia ventricosa* and *Pelvetia fastigiata* (Gutknecht, 1966; Allen, Jacobsen, Joaquin & Jaffe, 1972), but on the whole it seems more likely that they have 'normal' cytoplasm and follow alternative (3), using 'normal' methods of altering cytoplasmic osmotic strength. These 'normal' methods are to use non-ionic or zwitterionic substances for osmotica (Wyn Jones *et al.*, 1977), and it seems likely that an amino acid such as proline, which accumulates in some species in response to short-term water stress, or polyols or sugars which also can accumulate in water stress, could be involved in the lowering of guard cell cytoplasmic osmotic potential. Perhaps the accumulation of sucrose with stomatal opening, which some workers observe in guard cells (several species) although others do not (summarized by Outlaw & Manchester, 1979) could suggest sucrose as a cytoplasmic osmoticum. We can easily see the advantage of using a compound such as sucrose, which can readily be synthesized and just as readily removed, for this purpose; its use as a cytoplasmic osmoticum could also encourage the disappearance of starch with stomatal opening. Perhaps, thus, the oldest of the hypotheses (Kohl, 1895) might in due course reappear as an explanation for cytoplasmic osmotic control. There must in any case be some fast and effective method of keeping the guard cells' cytoplasmic osmotic pressure the same as that of the vacuole during the rapid changes that take place in stomatal opening: to identify this method is a challenge that seems at present too difficult.

I thank D. Talbot for a literature search, C. A. Barmby for permission to use some of her data, and A. G. Kirby-Brown and A. Sales for assistance. Research of my own reported here was supported by the Australian Research Grants Committee and the University of Sydney.

References

Allaway, W. G. (1973). Accumulation of malate in guard cells of *Vicia faba* during stomatal opening. *Planta* **110**, 63–70.

Allaway, W. G. (1976). Influence of stomatal behaviour on long distance transport. In *Transport and Transfer Processes in Plants*, ed. I. F. Wardlaw & J. Passioura, pp. 295–311. New York: Academic Press.

Allaway, W. G. & Hsiao, T. C. (1973). Preparation of rolled epidermis of *Vicia faba* L. so that stomata are the only viable cells: analysis of guard cell potassium by flame photometry. *Australian Journal of Biological Sciences* **26**, 309–18.

Allaway, W. G. & Milthorpe, F. L. (1975). Structure and functioning of stomata. In *Water Deficits and Plant Growth*, ed. T. T. Kozlowski, vol. IV, pp. 57–102. New York: Academic Press.

Allen, R. D., Jacobsen, L., Joaquin, J. & Jaffe, L. J. (1972). Ionic concentrations in developing *Pelvetia* eggs. *Developmental Biology* **27**, 538–45.

Barmby, C. A. (1979). Organic anions in stomatal opening in *Vicia faba*. B.Sc. Thesis, University of Sydney.

Bowling, D. J. F. (1976). Malate-switch hypothesis to explain the action of stomata. *Nature* **262**, 393–4.

Bowling, D. J. F. (1977). Mechanism of stomatal movement. *Nature* **266**, 282.

Das, V. S. R., Rao, I. M. & Raghavendra, A. S. (1977). Mechanism of stomatal movement. *Nature* **266**, 282.

Dayanandan, P. & Kaufman, P. B. (1973). Stomata in *Equisetum*. *Canadian Journal of Botany* **51**, 1555–64.

Dittrich, P. & Mayer, M. (1978). Inhibition of stomatal opening during uptake of carbohydrates by guard cells in isolated epidermal tissues. *Planta* **139**, 167–70.

Dittrich, P., Mayer, M. & Meusel, M. (1979). Proton-stimulated opening of stomata in relation to chloride uptake by guard cells. *Planta* **144**, 305–9.

Dittrich, P. & Raschke, K. (1977a). Malate metabolism in isolated epidermis of *Commelina communis* L. *Planta* **134**, 77–81.

Dittrich, P. & Raschke, K. (1977b). Uptake and metabolism of carbohydrates by epidermal tissue. *Planta* **134**, 83–90.

Fischer, R. A. (1968). Stomatal opening: role of potassium uptake by guard cells. *Science* **160**, 784–5.

Fischer, R. A. (1973). The relationship of stomatal aperture and guard-cell turgor pressure in *Vicia faba*. *Journal of Experimental Botany* **24**, 387–99.

Fischer, R. A. & Hsiao, T. C. (1968). Stomatal opening in isolated epidermal strips of *Vicia faba*. II. Response to KCl concentration and the role of potassium absorption. *Plant Physiology* **43**, 1953–8.

Greenway, H. & Sims, A. P. (1974). Effects of high concentrations of KCl and NaCl on responses of malate dehydrogenase (decarboxylating) to malate and various inhibitors. *Australian Journal of Plant Physiology* **1**, 15–30.

Gutknecht, J. (1966). Sodium, potassium and chloride transport and membrane potentials in *Valonia ventricosa*. *Biological Bulletin* **130**, 331–4.

Heller, F. O. & Resch, A. (1967). Funktionell bedingter Strukturwechsel der Zellkerne in den Schliesszellen von *Vicia faba*. *Planta* **75**, 243–52.

Hope, A. B. & Walker, N. A. (1975). *The Physiology of Giant Algal Cells*. Cambridge University Press.

Humbert, C. & Guyot, M. (1972). Modifications ultrastructurales des cellules stomatiques d'*Anemia rotundifolia* Schrad. *Comptes Rendus, Académie des Sciences Paris* **274**, 380–2.

Humbert, C., Louguet, P. & Guyot, M. (1975). Etude ultrastructurale comparée des cellules stomatiques de *Pelargonium* × *hortorum* en relation avec un état d'ouverture ou de fermeture des stomates physiologiquement défini. *Comptes Rendus, Académie des Sciences Paris* **280**, 1373–6.

Humble, G. D. & Hsiao, T. C. (1969). Specific requirement of potassium for light-activated opening of stomata in epidermal strips. *Plant Physiology* **44**, 230–4.

Humble, G. D. & Raschke, K. (1971). Stomatal opening quantitatively related to potassium transport. Evidence from electron probe analysis. *Plant Physiology* **48**, 447–53.

Jeschke, W. D. & Stelter, W. (1976). Measurement of longitudinal ion profiles in single roots of *Hordeum* and *Atriplex* by use of flameless atomic absorption spectroscopy. *Planta* **128**, 107–12.

Kirst, G. O. (1977). Ion composition of unicellular marine and freshwater algae, with special reference to *Platymonas subcordiformis* cultivated in media with different osmotic strengths. *Oecologia* **28**, 177–89.

Kohl, F. G. (1895). Über Assimilationsenergie und Spaltöffnungsmechanik. *Botanisches Centralblatt* **64**, 109–10.

Mansfield, T. A. & Jones, R. J. (1971). Effects of abscisic acid on potassium uptake and starch content of stomatal guard cells. *Planta* **101**, 147–58.

Milthorpe, F. L., Thorpe, N. & Willmer, C. M. (1979). Stomatal metabolism – a current assessment of its features in *Commelina*. In *Structure, function and ecology of stomata*, ed. D. D. Sen, D. D. Chawan & R. P. Bansal, pp. 121–42. Dehra Dun: Bishen Singh & Mahendra Pal Singh.

Moody, W. & Zeiger, E. (1978). Electrophysiological properties of onion guard cells. *Planta* **139**, 159–65.

Ogawa, T., Ishikawa, H., Shimada, K. & Shibata, K. (1978). Synergistic action of red and blue light and action spectra for malate formation in guard cells of *Vicia faba* L. *Planta* **142**, 61–5.

Osmond, C. B. (1978). Crassulacean acid metabolism: a curiosity in context. *Annual Review of Plant Physiology* **29**, 379–414.

Outlaw, W. H., Jr, & Kennedy, J. (1978). Enzymic and substrate basis for the anaplerotic step in guard cells. *Plant Physiology* **62**, 648–52.

Outlaw, W. H., Jr, & Lowry, O. H. (1977). Organic acid and potassium accumulation in guard cells during stomatal opening. *Proceedings of the National Academy of Sciences, USA* **74**, 4434–8.

Outlaw, W. H., Jr, & Manchester, J. (1979). Guard cell starch concentration quantitatively related to stomatal aperture. *Plant Physiology* **64**, 79–82.

Outlaw, W. H., Jr, Manchester, J. & Di Camelli, C. A. (1979*a*). Histochemical approach to properties of *Vicia faba* guard cell phosphoenolpyruvate carboxylase. *Plant Physiology* **64**, 269–72.

Outlaw, W. H. Jr, Manchester, J., Di Camelli, C. A., Randall, D. D., Rapp, B. & Veith, G. M. (1979*b*). Photosynthetic carbon reduction pathway absent in chloroplasts of *Vicia faba* L. guard cells. *Proceedings of the National Academy of Sciences, USA* **76**, 6371–5.

Palevitz, B. A. & Hepler, P. K. (1976). Cellulose microfibril orientation and cell shaping in developing guard cells of *Allium*: The role of microtubules and ion accumulation. *Planta* **132**, 71–93.

Pallaghy, C. K. (1968). Electrophysiological studies in guard cells of tobacco. *Planta* **80**, 147–53.

Pallaghy, C. K. (1971). Stomatal movement and potassium transport in epidermal strips of *Zea mays*: The effect of CO_2. *Planta* **101**, 287–95.

Pallaghy, C. K. & Fischer, R. A. (1974). Metabolic aspects of stomatal opening and ion accumulation by guard cells in *Vicia faba*. *Zeitschrift für Pflanzenphysiologie* **71**, 332–44.

Pallas, J. E., Jr, & Wright, B. G. (1973). Organic acid changes in the epidermis of *Vicia faba* and their implication in stomatal movement. *Plant Physiology* **51**, 588–90.

Pearson, C. J. & Milthorpe, F. L. (1974). Structure, carbon dioxide fixation and metabolism of stomata. *Australian Journal of Plant Physiology* **1**, 221–36.

Penny, M. G. & Bowling, D. J. F. (1974). A study of potassium gradients in the epidermis of intact leaves of *Commelina communis* L. in relation to stomatal opening. *Planta* **119**, 17–25.

Penny, M. G. & Bowling, D. J. F. (1975). Direct determination of pH in the stomatal complex of *Commelina*. *Planta* **122**, 209–12.

Penny, M. G., Kelday, L. S. & Bowling, D. J. F. (1976). Active chloride transport in the leaf epidermis of *Commelina communis* in relation to stomatal activity. *Planta* **130**, 291–4.

Pitman, M. G. & Saddler, H. D. W. (1967). Active sodium and potassium transport in cells of barley roots. *Proceedings of the National Academy of Sciences, USA* **57**, 44–9.

Prosser, C. L. (1973). *Comparative Animal Physiology*. 3rd edn. Philadelphia: Saunders.

Raghavendra, A. S., Rao, I. M. & Das, V. S. R. (1976). Shrinkage of guard cell chloroplasts in relation to stomatal opening in *Commelina benghalensis* L. *Annals of Botany* **40**, 899–901.

Raschke, K. & Dittrich, P. (1977). [^{14}C] carbon-dioxide fixation by isolated leaf epidermis with stomata closed or open. *Planta* **134**, 69–75.

Raschke, K. & Fellows, M. P. (1971). Stomatal movement in *Zea mays*: shuttle of potassium and chloride between guard cells and subsidiary cells. *Planta* **101**, 296–316.

Raschke, K. & Humble, G. D. (1973). No uptake of anions required by opening stomata of *Vicia faba*: guard cells release hydrogen ions. *Planta* **115**, 47–57.

Raschke, K. & Schnabl, H. (1978). Availability of chloride affects the balance between potassium chloride and potassium malate in guard cells of *Vicia faba* L. *Plant Physiology* **62**, 84–7.

Sawhney, B. L. & Zelitch, I. (1969). Direct determination of potassium ion accumulation in guard cells in relation to stomatal opening in light. *Plant Physiology* **44**, 1350–4.

Schnabl, H. (1978). The effect of Cl$^-$ upon the sensitivity of starch-containing and starch-deficient stomata and guard cell protoplasts towards potassium ions, fusioccin and abscisic acid. *Planta* **144**, 95–100.

Schnabl, H. & Ziegler, H. (1977). The mechanism of stomatal movement in *Allium cepa* L. *Planta* **136**, 37–43.

Thorpe, N., Brady, C. J. & Milthorpe, F. L. (1978). Stomatal metabolism: primary carboxylation and enzyme activities. *Australian Journal of Plant Physiology* **5**, 485–93.

Thorpe, N. & Milthorpe, F. L. (1977). Stomatal metabolism: CO_2 fixation and respiration. *Australian Journal of Plant Physiology* **4**, 611–21.

Thorpe, N. & Milthorpe, F. L. (1979). Stomatal metabolism: carbon dioxide fixation and labelling patterns during stomatal movement in *Commelina cyanea*. *Australian Journal of Plant Physiology* **6**, 409–16.

Travis, A. J. & Mansfield, T. A. (1977). Studies of malate formation in 'isolated' guard cells. *New Phytologist* **78**, 541–6.

Van Kirk, C. A. & Raschke, K. (1978*a*). Presence of chloride reduces malate production in epidermis during stomatal opening. *Plant Physiology* **61**, 361–4.

Van Kirk, C. A. & Raschke, K. (1978*b*). Release of malate from epidermal strips during stomatal closure. *Plant Physiology* **61**, 474–5.

Willmer, C. M. & Dittrich, P. (1974). Carbon dioxide fixation by epidermal and mesophyll tissues of *Tulipa* and *Commelina*. *Planta* **117**, 123–32.

Willmer, C. M. & Pallas, J. E., Jr, (1973). A survey of stomatal movements and associated potassium fluxes in the plant kingdom. *Canadian Journal of Botany* **51**, 37–42.

Willmer, C. M., Pallas, J. E., Jr, & Black, C. C. (1973). Carbon dioxide metabolism in leaf epidermal tissue. *Plant Physiology* **52**, 448–52.

Willmer, C. M. & Rutter, J. C. (1977). Guard cell malic acid metabolism during stomatal movements. *Nature* **269**, 327–8.

Willmer, C. M., Thorpe, N., Rutter, J. C. & Milthorpe, F. L. (1978). Stomatal metabolism: carbon dioxide fixation in attached and detached epidermis of *Commelina*. *Australian Journal of Plant Physiology* **5**, 767–78.

Wyn Jones, R. G., Storey, R., Leigh, R. A., Ahmad, N. & Pollard, A. (1977). A hypothesis on cytoplasmic osmoregulation. In *Regulation of Cell Membrane Activities in Plants*, ed. E. Marrè and O. Ciferri, pp. 121–36. Amsterdam: Elsevier.

Zeiger, E., Moody, W., Hepler, P. & Varela, F. (1977). Light-sensitive membrane potentials in onion guard cells. *Nature* **270**, 270–1.

COLIN M. WILLMER

Guard cell metabolism

Introduction

In 1950 O. V. S. Heath mentioned the possibility of dark carboxylation reactions being involved in stomatal mechanisms (Heath, 1950). It was over 20 years later that real evidence emerged to support this view when Allaway (1973) found that malic acid levels in guard cells of *Vicia faba* L. increased as stomata opened and Willmer *et al.* (1973*a*, *b*) found high levels of phosphoenolpyruvate (PEP) carboxylase and other enzymes concerned in malate metabolism in epidermal tissue, probably in guard cells. Since then an ever-increasing volume of literature on guard cell metabolism has amassed. It is the intention in this chapter to review the literature and give an up-to-date assessment of our current knowledge about guard cell metabolism. Information on enzyme activities within guard cells and epidermal tissue and on $^{14}CO_2$ uptake kinetics and turnover patterns by epidermal tissue will be integrated in efforts to establish important pathways occurring in guard cells during stomatal movements.

Enzyme activities in guard cells and epidermal tissue

Information on guard cell metabolism has been obtained chiefly from experiments using conventional techniques such as feeding labelled metabolites to isolated epidermis and observing turnover patterns, or by studying enzymes from epidermal tissue. Recently, however, a technique for the biochemical analysis of guard cell pairs has been developed (Outlaw & Lowry, 1977; Outlaw & Kennedy, 1978) and, doubtless, more valuable information will arise from the use of this technique in the near future. Measurements of key enzymes in epidermal tissue or in individual cells of the epidermal layer have greatly assisted our understanding of guard cell metabolism. However, certain problems must be appreciated when measuring enzyme activities in epidermal tissue. One problem that may occur is contamination of the epidermal tissue with underlying mesophyll. Assays for enzymes within such contaminated epidermal tissue can lead to misleading results about enzyme location and activities (see Raschke & Dittrich, 1977). It is essential, therefore, to establish that the epidermal tissue is uncontaminated by other tissues. Another problem that can occur is that, in some species and for certain enzymes, crude epidermal extracts and some extracts after being passed through a G-25 Sephadex column, may contain enzyme inhibitors. For example, equal volumes of a crude extract from the epidermis of *V. faba* caused a 30% reduction in activity of NADP specific malic enzyme in a *Sorghum* leaf extract (Willmer, unpublished observations) and small additions of a crude epidermal extract from *Pelargonium zonale* Ait. caused complete inhibition of NAD-specific malate dehydrogenase in a crude extract from maize leaves (Jamieson & Willmer, unpublished observations).

Table 1 gives a list of key enzymes which have been found in epidermal extracts from *Commelina cyanea* R.Br. by Thorpe, Brady & Milthorpe (1978). These activities are fairly representative of what other workers have found in other species (Willmer *et al.*, 1973*a, b*; Das & Raghavendra, 1974). A general feature is the presence of enzymes typical of the primary carbon dioxide fixation and decarboxylation pathways found in C_4 and CAM plants. Enzyme activities have been detected at levels high enough to enable the following series of events to occur at rates equal to or greater than rates of carbon dioxide fixation found in crude or partially purified leaf extracts from C_4 plants.

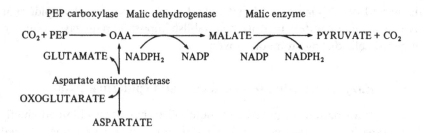

Normally there is little difficulty in detecting high levels of PEP carboxylase (E.C. 4.1.1.31) and NAD malate dehydrogenase (E.C.1.1.1.37) activity in crude epidermal tissue extracts from all species so far tested. However, it is our experience that NADP malate dehydrogenase (E.C. 1.1.1.82) and NADP and NAD malic enzymes (E.C. 1.1.1.40 and 1.1.1.38, respectively) are more difficult to detect, being at much lower activities than NAD malate dehydrogenase or PEP carboxylase. NADP-specific malic enzyme predominates in epidermal tissue although Thorpe *et al.* (1978) also detected low levels of NAD-specific malic enzyme. These authors also found much greater activities of NADP malic enzyme in the presence of Mn^{2+} ions rather than Mg^{2+} ions which activate the enzyme in C_4 leaves.

^{14}C-labelling studies

Metabolic pathways suggested by the enzyme studies are supported by results from ^{14}C labelling experiments. The original labelling experiments of Willmer & Dittrich (1974) in which epidermal tissue of *Commelina diffusa* L. was suspended in a $NaH^{14}CO_3$ solution serve to illustrate the usual turnover pattern of labelled compounds (Fig. 1). ^{14}C label is chiefly located in malate which remains heavily labelled over short-term labelling experiments, with lesser amounts in aspartate, sugars and sugar phosphates. A small amount of label may also appear in amino acids (other than aspartate) and intermediates of the tricarboxylic acid cycle (other than malate). Later studies in which epidermis from various species was exposed to $^{14}CO_2$ or $NaH^{14}CO_3$ for short-term periods (of the order of a few minutes) essentially confirmed this picture (Raschke & Dittrich, 1977; Willmer, Thorpe, Rutter & Milthorpe, 1978; Milthorpe, Thorpe & Willmer, 1979).

In epidermis of *Allium cepa* L. fed with $^{14}CO_2$ Schnabl (1977) also found that organic acids and amino acids became labelled but not sugars, sugar phosphates or 3-phosphoglyceric acid.

Normally malate is the major labelled compound in short-term $^{14}CO_2$ feeding experiments but in epidermis of *C. cyanea* a considerable amount of ^{14}C label also appears in aspartate (Thorpe & Milthorpe, 1977; Thorpe *et al.*,

Table 1. *Enzyme activities in epidermal tissue of* Commelina cyanea. *Data taken from Thorpe, Brady & Milthorpe (1978)*

Enzyme	μmole of substrate transformed per h	
	per mg protein	per mg chlorophyll
RuBP carboxylase	0.2	22
PEP carboxylase	5.2	712
Malic enzyme (NAD) Mg^{2+}	N.D.	N.D.
Malic enzyme (NAD) Mn^{2+}	2.4	342
Malic enzyme (NADP) Mg^{2+}	15.5	2204
Malic enzyme (NADP) Mn^{2+}	45.8	6627
Malic dehydrogenase (NAD)	182.3	28 700
Malic dehydrogenase (NADP)	10.6	1537
Pyruvate P_i dikinase	N.D.	N.D.
PEP carboxykinase	N.D.	N.D.
Asp aminotransferase	52.8	7482
Ala aminotransferase	3.3	413

N.D., not detectable.

1978). In epidermis of *Commelina communis* L. and *C. diffusa*, however, malate is labelled most heavily and if uniformly labelled (U) ^{14}C aspartate is fed to epidermal tissue of *C. communis* much of the label is rapidly found in malate (Willmer, unpublished observations). Since high levels of NADP-specific malic enzyme have been detected in the epidermis of all three *Commelina* species and malate is the dominantly labelled compound, their guard cells may be considered 'malate' formers (of the NADP malic enzyme type) rather than 'aspartate' formers. Nevertheless, Thorpe *et al.* (1978) found some NAD malic enzyme, aspartate aminotransferase (E.C. 2.6.1.1) and alanine aminotransferase (E.C. 2.6.1.2) activity in epidermis of *C. cyanea* suggesting that the guard cells of this species operate, to some extent, as 'aspartate' formers of the NAD malic enzyme type.

Fig. 1. The distribution of ^{14}C amongst metabolites after exposure of epidermal tissue of *Commelina diffusa* to NaH[14]CO$_3$ in the light (*a*) and the dark (*b*). Open circles, malate; open triangles, aspartate; filled circles, sugar phosphates + PGA; filled triangles, sucrose; crosses, citrate, open squares, glutamate. Adapted from Willmer & Dittrich (1974).

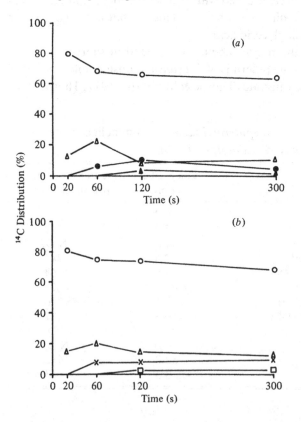

It has been speculated that the pyruvate formed in the decarboxylation reaction was used to regenerate PEP and that guard cells performed some sort of CAM or, in conjunction with other cell types, C_4 photosynthesis (Willmer & Dittrich, 1974). However, the ^{14}C turnover patterns are unlike those observed in CAM or C_4 plants in that labelled malate turns over very slowly and the amount of label in sugars and sugar phosphates remains low or none is present.

There is much evidence showing that starch levels in guard cells decrease (see Meidner & Mansfield, 1968, pp. 73–6; Outlaw & Manchester, 1979) while malate levels increase (Allaway, 1973; Pearson & Milthorpe, 1974; Travis & Mansfield, 1977) during stomatal opening, the reverse situation occurring upon closure. Thus, in order to understand the fate of pyruvate, and to follow the flow of carbon between malate and starch, experiments were conducted in which labelled substances were applied to epidermal tissue with stomata at different phases of movements.

Dittrich & Raschke (1977a) found that when 4-^{14}C malic acid was fed to epidermal strips of *C. communis* in the light or dark for two hours (stomata remained closed), little radioactivity could be detected in sugars. In the presence of U-^{14}C malate, however, more label accumulated in sugars after incubation for two hours. Unfortunately, the U-^{14}C malate was supplied to epidermal tissue at pH 3.3, a pH which slowly kills all the cells, while 4-^{14}C malate was supplied to the epidermal tissue at pH 5.7 so that it is difficult to interpret the experiment precisely. Nevertheless, the data suggest that when stomata are closed, at least some of the carbon from malate is channelled into sugars, presumably *via* gluconeogenesis (see Fig. 3), and this process involves the decarboxylation of malate.

In a similar experiment conducted by Willmer & Rutter (1977) U-^{14}C malic acid was fed to epidermal strips of *C. communis* when stomata were opening or closing in a buffered medium at pH 5.5. They found that the rates of malate uptake and ^{14}C incorporation into starch were highest when stomata were closing although label also became incorporated into starch when stomata were opening. These experiments suggest that upon decarboxylation of malate, at least some of the pyruvate formed is channelled into the gluconeogenic pathway, ultimately finding its way into starch (see Fig. 3). Willmer & Rutter (1977) also concluded that there was a dynamic flux of carbon between malic acid and starch with the direction of starch production outpacing its breakdown when stomata were closing.

Pyruvate cannot be directly channelled into gluconeogenesis since one step in the glycolytic sequence, PEP to pyruvate, is essentially irreversible. Steps at the PEP stage and beyond are reversible in the glycolytic/ gluconeogenic pathway. Thus, a search for PEP synthesizing enzymes has been made. Das & Raghavendra (1974) detected pyruvate P_i dikinase (E.C. 2.7.9.1) in epidermal tissue of *Commelina benghalensis* L. and *Tridax procumbens* L., enabling pyruvate to be first converted to PEP before being channelled into starch via gluconeogenesis. Pyruvate P_i dikinase was not detected in epidermal tissue of *C. cynea* (Thorpe *et al.*, 1978) or *C. communis* (Willmer, unpublished observations) however, and thus, the conversion of pyruvate to PEP catalysed by pyruvate P_i dikinase does not appear to be a common feature of epidermal tissue from all species.

Similarly, PEP carboxykinase (E.C. 4.1.1.32), another enzyme found in some C_4 species and which catalyses the synthesis of PEP from OAA and ATP, could not be detected in epidermal tissue of *C. cyanea* (Thorpe *et al.*, 1978) or *C. communis* (Willmer, unpublished observations).

Pyruvate can also be directly channelled into the tricarboxylic acid cycle with subsequent production of ATP and reducing power which would be available for active ion transport and other energy consuming processes in the guard cell. The appearance of ^{14}C label in intermediates of the tricarboxylic acid cycle such as citrate, fumarate and succinate, after $^{14}CO_2$, ^{14}C malate or ^{14}C aspartate is fed to epidermal tissue suggests that this happens.

A number of studies have also been made in which $^{14}CO_2$ was fed to epidermal tissue and stomatal apertures measured. Raschke & Dittrich (1977) found that when epidermes of tulip and *C. communis* were exposed to $^{14}CO_2$, the labelling patterns did not differ when stomata were open or closed and the main labelled products were malate and aspartate.

In a more detailed study of carbon dioxide fixation by epidermal tissue of *C. cyanea* with stomata at different phases of movement, Thorpe, Willmer & Milthorpe (1979) found that *rates* of CO_2 fixation were about three times higher when the stomata were opening or closing than when they were either open or closed. Also, the proportion of label in the various products was similar during the opening, open and closed phases but, when closing, a higher proportion of label was diverted to sugars and sugar phosphates.

In the study of Thorpe *et al.* (1979) epidermal tissue was exposed to $^{14}CO_2$ for short periods of time (up to a few minutes) to determine the fixation kinetics and turnover patterns. In a recent study (Freer-Smith & Willmer, unpublished observations) epidermal tissue of *C. communis* with stomata opening or closing was exposed to $^{14}CO_2$ for long time periods (up to 2 h) and the turnover patterns and uptake kinetics studied. Results showed a long-term turnover of ^{14}C from malate to sugars during opening while, during closure, movement of label between metabolites was less obvious with considerable amounts of label initially in sugars, sugar phosphates and malate (Fig. 2). Starch became heavily labelled during both opening and closing phases. There was also a marginally higher rate of carbon dioxide incorporation when stomata were opening than when they were closing. The results of this study suggest that there may be a pool of sugar (mainly sucrose) in epidermal tissue, possibly in guard cells, from which carbon can be channelled into starch synthesis (when stomata are closing) or malate synthesis (when stomata are opening; (Fig. 3).

The Calvin cycle and guard cells

A contentious issue is whether ribulose bisphosphate (RuBP) carboxylase (E.C. 4.1.1.39) is present in guard cells and, if so, whether the Calvin cycle is operable. RuBP carboxylase activity has been found in the epidermis of a number of species by various workers: Willmer *et al.* (1973a) detected the enzyme in the epidermis of tulip and *C. communis*; Thorpe *et al.* (1978) detected low levels of activity in the epidermis of *C. cyanea* and Jamieson & Willmer (unpublished observations) also found activity in leek and onion epidermis. Raschke & Dittrich (1977), however, believe that, at

Fig. 2. The distribution of ^{14}C amongst metabolites present within epidermal tissue of *C. communis* exposed to $^{14}CO_2$ during stomatal opening (*a*) and closing (*b*). Open circles, malate; open triangles, aspartate; filled circles, sugar phosphates; filled triangles, sucrose; crosses, others.

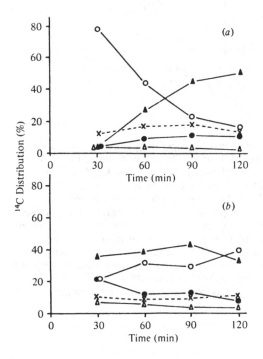

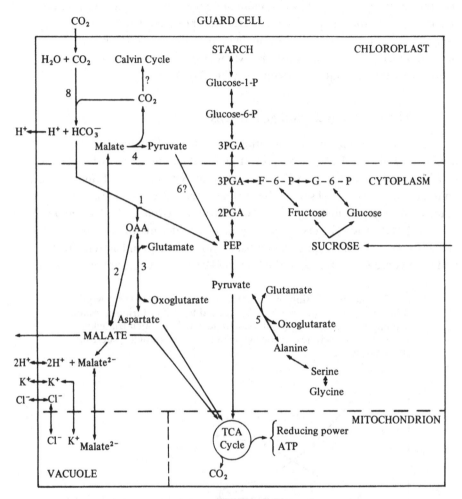

Fig. 3. Major pathways which may be operating in guard cells during
stomatal movements (after Milthorpe, Thorpe & Willmer, 1979). The
numbers represent enzymes catalysing the reactions. KEY: 1, PEP
carboxylase; 2, NADP (NAD) malate dehydrogenase; 3, aspartate
aminotransferase; 4, NADP (NAD) malic enzyme; 5, alanine
aminotransferase; 6, pyruvate phosphate dikinase (this enzyme has, so
far, not been detected in epidermal tissue); 7, phosphorylase; 8, carbonic
anhydrase. Assuming cytoplasmic pH remains constant around 7 then
malate will be about 99% in the mal $^{2-}$ form. However, vacuolar pHs
change and values around 5 have been recorded when stomata close. At
pH 5.0, malate is ionized into approximately 50% H mal^{1-} and 50%
mal^{2-} with negligible H_2 mal. Dissociation constants indicate little
binding of malate and K^+.

least in *C. communis* and possibly tulip, the Calvin cycle is absent and suggest that detection of RuBP carboxylase in the epidermis is principally the result of contamination by the mesophyll. Raschke & Dittrich (1977) found that clean epidermal strips of *C. communis* could not catalyse the conversion of exogeneously supplied R-5-P to RuBP although homogenate from the strips could convert exogenous RuBP to 3PGA at a low rate. Thus, their conclusion was that phosphoribulose kinase (E.C. 2.7.1.19) was absent although low levels of RuBP carboxylase were present in epidermis of *C. communis*.

Schnabl (1977) also concluded from ^{14}C-labelling studies that the Calvin cycle does not operate in the epidermis of *Allium cepa* since no radioactivity was found in PGA or sugar phosphates.

Recent work by Outlaw *et al.* (1979), using enzyme assays on single guard cell pairs, essentially supports the results of Dittrich & Raschke (1977). Outlaw *et al.* (1979) could not detect RuBP carboxylase in guard cells of *V. faba* or tobacco and only trace amounts were found in epidermal cells of the lower epidermis. Very low or insignificant levels of phosphoribulokinase and glyceraldehyde-P dehydrogenase (E.C. 1.2.1.13) were also found in guard cells and epidermal cells. Additionally, immunoelectrophoresis studies did not detect RuBP carboxylase in extracts from guard cell protoplasts of *V. faba*.

In certain species of *Paphiopedilum* (Nelson & Mayo, 1975; Rutter & Willmer, 1979) and *Pelargonium zonale* cv. Chelsea Gem (Jamieson & Willmer, unpublished observations), the Calvin cycle is not present in guard cells since chloroplasts are absent although the stomata are functional. Thus, in some species, at least, the Calvin cycle is not essential for the functioning of stomata. Guard cell chloroplasts, where present, may however, be an important source of ATP and reducing power for ion-transport and other processes essential to the functioning of stomata. Also, as already indicated, the ^{14}C turnover patterns (see Fig. 1) are not typical of C_4 or CAM plants with very little build up of label in 3-PGA and sugar phosphates, suggesting that the importance of the Calvin cycle in guard cell metabolism may be negligible.

Fluxes of metabolites between guard cells and other cells

If it is a general feature that the Calvin cycle is absent from guard cells and other cell types within the epidermis then it is very unlikely there will be a net fixation of carbon dioxide by this tissue. The C_4 pathway is not autocatalytic and, unless there are alternatives to the Calvin cycle, this remains the only possible autocatalytic pathway for carbon dioxide fixation (Kelly & Latzko,

1976). For the epidermal tissue to survive there must, therefore, be import of carbon from the rest of the leaf. Evidence to suggest that this occurs is available. First, rates of carbon dioxide fixation are much higher in attached than detached epidermis in all species studied so far (Pearson & Milthorpe, 1974; Thorpe & Milthorpe, 1977; Dittrich & Raschke, 1977b; Willmer et al., 1978). Second, autoradiographs of attached epidermis fed $^{14}CO_2$ usually show label localized throughout the epidermal tissue while in detached epidermis exposed to $^{14}CO_2$, label is generally located over the stomata.

There are exceptions to these observations, however. For instance, detached epidermis of *Paphiopedilum venustum* and *P. harriseanum* Hort., species which do not possess guard cell chloroplasts, shows low CO_2-fixing ability with no concentration of ^{14}C label over the stomata. Similarly, detached epidermis of *Pelargonium zonale* cv. Chelsea Gem, taken from leaves with non-chlorophyllous guard cells, does not show an accumulation of label over the stomata; instead, glandular, multicellular trichomes became heavily labelled and, therefore, appear to be the most active cells at fixing carbon dioxide. In *Polypodium vulgare* L. (a fern) label can be located throughout both attached and detached epidermis with slightly heavier labelling over the stomata. In *P. vulgare*, however, not only are the guard cells full of chloroplasts but the epidermal cells also contain an abundance of well-developed chloroplasts which may be able to fix carbon dioxide.

A third line of evidence to suggest metabolite transport from the mesophyll to the epidermis was obtained by Dittrich & Raschke (1977b). They fed a leaf, with part of the lower epidermis missing, $^{14}CO_2$ for 10 min. After a rinse in distilled water the piece of epidermis was replaced on the peeled surface and left for 3 h. Radioactivity was then detected in the previously peeled epidermis. Unfortunately errors of interpretation can arise from this sort of experiment. For example, mesophyll cells may be ruptured in peeling off the epidermis from the leaf so that when the epidermis is replaced photosynthate may leak from the damaged cells to contaminate the epidermis. Moreover, when similar experiments were conducted with the epidermis replaced by a similar-sized piece of moist filter paper, label could also be detected in the filter paper (Jamieson & Willmer, unpublished observations).

A fourth and more convincing line of evidence for metabolite transport from the mesophyll to the epidermis was obtained by Outlaw & Fisher (1975) and Outlaw, Fisher & Christy (1975). After feeding a pulse of $^{14}CO_2$ to a leaf of *V. faba* they followed the partitioning of label between various tissues with an elegant freeze-substitution technique, and discovered that label eventually reached the epidermal tissue.

Another piece of evidence comes from measurements of the carbon isotope discrimination of epidermal and mesophyll tissues from various C_3 species. Although ^{13}C values for epidermal tissue were marginally less negative than values for mesophyll tissue, values for both tissues were typical of C_3 plants (Willmer & Firth, 1979). This suggests not only that metabolites are transported from the mesophyll to the epidermis but also that PEP carboxylase does not make a large contribution to the metabolism of epidermal tissue.

It is important to know if metabolites move between the guard cells and neighbouring cells during the functioning of stomata. In early experiments in which epidermal strips were incubated in $NaH^{14}CO_3$ solutions the possibility of either passive or active efflux of labelled metabolites from the tissue into the medium was not considered. Later work showed that metabolites did leave the epidermal tissue and entered the incubating medium. Such loss of ^{14}C labelled metabolites makes the kinetics and turnover patterns of carbon dioxide assimilation in epidermal tissue more difficult to interpret. Indeed, loss of metabolites may account, in part, for the low assimilation rates of carbon dioxide which are observed in isolated epidermis compared to epidermis located *in situ* within the leaves.

Dittrich & Raschke (1977a) observed an exit of labelled malate alone from epidermal strips of *C. communis* when stomata were induced to close rapidly with ABA (5 min for full closure) or more slowly by placing the strips on water (about 30 min for full closure). They concluded from this study that a large proportion of malic acid synthesized in the guard cells was exported from these cells during stomatal closure. This view is supported by experiments of Van Kirk & Raschke (1978a) which showed that, when stomata closed, guard cells in epidermal strips of *C. communis* and *V. faba* released into the incubating medium a substantial amount of malate only.

Others (Thorpe & Milthorpe, 1977; Thorpe *et al.*, 1979; Freer-Smith & Willmer, unpublished observations), however, have found that numerous labelled substances 'leaked' from epidermal tissue at all stages of stomatal movements. In *C. cyanea* Thorpe & Milthorpe (1977) found no selective retention of labelled compounds within the epidermis. In later work it was discovered that the amount of labelled compounds 'leaked' into the incubation medium varied with the phase of stomatal movement and that the tissue lost less malate when stomata were opening than at other times (Thorpe *et al.*, 1979). In *C. communis*, although label could be detected in various intermediates of the tricarboxylic acid cycle, and in amino acids and sugars, the label was predominantly located in malic acid when the stomata were opening or closing (Freer-Smith & Willmer, unpublished observations: Table 2). Although these data favour the view that malate is the most important substance lost from guard cells, the additional finding that substantial amounts of malate are lost when stomata are both closing *and* opening, is difficult to interpret. It would appear more logical if loss of solute occurred only during the closing phase when guard cell osmotic potentials were increasing.

Freer-Smith & Willmer used 10^{-8} M ABA in 0.1 mM $CaCl_2$ to cause stomatal closure and 100 mM NaCl to stimulate opening. These media were used to obtain controlled and repeatable opening and closing kinetics but their use does create problems in interpreting the results. For example, Van Kirk & Raschke (1978*b*) found much more malate was formed in detached epidermis of *V. faba* when chloride ions were absent from the incubating medium. Also, the different incubating media used to cause opening or closing movements may affect membrane permeabilities to different extents.

The metabolism of onion guard cells

The stomatal mechanism in certain members of the Liliaceae, Iridaceae and Amaryllidaceae deserves special attention since their guard cells lack starch (Steinberger, 1922; Allaway & Setterfield, 1972; Schnabl, 1977; Schnabl & Ziegler, 1977). Since starch is considered to play an integral part in the stomatal mechanism by supplying part of the carbon skeleton for malate synthesis it has been suggested that fructosans or some other soluble polysaccharide serves this purpose in starch-free guard cells. However, Schnabl (1977) found that, although a water-soluble, mucilaginous polysaccharide could be isolated from epidermal tissue of *Allium cepa*, no fructosans were present. It was concluded that the mucilaginous polysaccharides did not serve as a carbon source for malate synthesis (Schnabl,

Table 2. *Percentage distribution of labelled metabolites 'leaked' from epidermal strips of* Commelina communis *after a 60 min stomatal opening or closing period. (Opening medium, 100 mM NaCl, pH 7.9; closing medium, 10^{-8} M ABA in 0.1 mM $CaCl_2$, pH 7.8. Each medium contained 0.8 mM NaH $^{14}CO_3$)*

Compound	Closing	Opening
	% distribution	
Malate	91.7	88.7
Sucrose	2.8	2.3
Aspartate	1.0	0.6
Citrate	N.D.	4.8
Fumarate	N.D.	1.6
Others	4.5	2.0

N.D., non-detectable.

1977) and that there were no other water-soluble polysaccharides present in *A. cepa* epidermis (Schnabl & Ziegler, 1977). Schnabl & Ziegler (1977) also found high malate concentrations in epidermis of *A. cepa* although the concentrations did not change as stomatal apertures changed. Additionally, they found an absolute chloride-ion requirement for potassium-ion uptake and stomatal opening in epidermal strips of *A. cepa* and, in 'Cl⁻ -watered plants, chloride ions could compensate fully for the potassium charges in the guard cells. It was concluded, therefore, that malate anions present in *A. cepa* guard cells play little or no part in maintaining charge balance when potassium ions enter guard cells. More recently, however, Schnabl (1978) discovered that high chloride levels caused malate anions to disappear from guard cell protoplasts of *A. cepa* and it was suggested that, contrary to earlier views, malate may have a role to play in the stomatal mechanism serving as a 'proton-primer' (see Dittrich, Mayer & Meusel, 1979). In support of the view that malate plays an important part in the stomatal mechanism of *Allium* are the findings of high levels of enzymes concerned in malate metabolism (PEP carboxylase, NADP malic enzyme and NADP malate dehydrogenase) in *A. cepa* and *A. porrum* epidermis (Jamieson & Willmer, unpublished observations). The source of carbon for malate synthesis and the sink for carbon upon malate breakdown still remain obscure.

Summary

Fig. 3 attempts to outline the most likely pathways of carbon flow during stomatal movements and summarizes important points mentioned in the earlier part of the text. It is generally accepted that during stomatal opening malate levels increase in the guard cells, primarily as a result of PEP carboxylase activity, with concomitant decreases in starch levels (in those species possessing guard cell starch), the reverse situation occurring during stomatal closure. The pathways of carbon flow between malate and starch during opening and closing are less clear but it is likely that, upon opening, starch is hydrolysed and carbon leaves the chloroplasts of the guard cell as triose phosphates or 3-PGA, eventually ending up in PEP and then malic acid. During stomatal closure, essentially the reverse pathway occurs: malate is decarboxylated and carbon from the resulting pyruvate is channelled into starch synthesis, presumably *via* gluconeogenesis. Malate and/or pyruvate can also be passed directly into the tricarboxylic acid cycle and be respired. Carbon dioxide resulting from the tricarboxylic acid cycle and from malic enzyme activity may be refixed via PEP carboxylase activity. There may also be a shuttle of malate between the guard cells and neighbouring cells during stomatal movements. There is some evidence that there is a pool of sugar (mainly sucrose) located between the starch/malate interconversion which may control rates and, to some extent, the direction of carbon flow.

The weight of evidence suggests that the Calvin cycle is absent from guard cells or, at least, has little importance in the biochemistry of the stomatal mechanism. Import of carbon by the guard cells from the mesophyll also occurs although with large 'stores' of guard cell starch and the absence of plasmadesmata between mature guard cells and neighbouring cells, the speed and efficiency of this process may not be important.

References

Allaway, W. G. (1973). Accumulation of malate in guard cells of *Vicia faba* during stomatal opening. *Planta* **110**, 63–70.

Allaway, W. G. & Setterfield, G. (1972). Ultrastructural observations on guard cells of *Vicia faba* and *Allium porrum*. *Canadian Journal of Botany* **50**, 1405–13.

Das, V. S. R. & Raghavendra, A. S. (1974). Control of stomatal opening by pyruvate metabolism in light. *Indian Journal of Experimental Biology* **12**, 425–8.

Dittrich, P. Mayer, M. & Meusel, M. (1979). Proton-stimulated opening of stomata in relation to chloride uptake by guard cells. *Planta* **144**, 305–9.

Dittrich, P. & Raschke, K. (1977*a*). Malate metabolism in isolated epidermis of *Commelina communis* L. in relation to stomatal functioning. *Planta* **134**, 77–81.

Dittrich, P. & Raschke, K. (1977*b*). Uptake and metabolism of carbohydrates by epidermal tissue. *Planta* **134**, 83–90.

Heath, O. V. S. (1950). Studies in stomatal behaviour. V. The role of carbon dioxide in the light response of stomata. *Journal of Experimental Botany* **1**, 29–62.

Kelly, J. G., Latzko, E. & Gibbs, M. (1976). Regulatory aspects of photosynthetic carbon metabolism. *Annual Review of Plant Physiology* **27**, 181–205.

Meidner, H. & Mansfield, T. A. (1968). *Physiology of Stomata*. London: McGraw-Hill.

Milthorpe, F. L., Thorpe, N. & Willmer, C. M. (1979). Stomatal metabolism – a current assessment of its features in *Commelina*. In *Structure, Function and Ecology of Stomata*, D. D. Cmawan & R. P. Bansal, pp. 121–42. ed. D. N. Sen, Dehra Dun: Bishen Singh & Mahendra Pal Singh.

Nelson, S. D. & Mayo, J. M. (1975). The occurrence of functional non-chlorophyllous guard cells in *Paphiopedilum* spp. *Canadian Journal of Botany* **53**, 1–7.

Outlaw, W. H. & Fisher, D. B. (1975). Compartmentation in *Vicia faba* leaves. I. Kinetics of ^{14}C in the tissues following pulse labelling. *Plant Physiology* **62**, 648–52.

Outlaw, W. H. & Kennedy, J. (1978). Enzymic and substrate basis for the anaplerotic step in guard cells. *Plant Physiology* **62**, 648–52.

Outlaw, W. H. & Lowry, O. M. (1977). Organic acid and potassium accumulation in guard cells during stomatal opening. *Proceedings of the National Academy of Sciences of the USA* **74**, 4434.

Outlaw, W. H., Fisher, D. B. & Christy, A. W. (1975). Compartmentation in *Vicia faba* leaves. II. Kinetics of ^{14}C-sucrose redistribution among individual leaves following pulse labelling. *Plant Physiology* **55**, 704–11.

Outlaw, W. H. & Manchester, J. (1979). Guard cell starch concentration quantitatively related to stomatal aperture. *Plant Physiology* **64**, 79–82.

Outlaw, W. H., Manchester, J., Dicamelli, C. A., Randall, D. D., Rapp, B. & Veith, G. M. (1979). Photosynthetic carbon reduction pathway absent in chloroplasts of *Vicia faba* guard cells. *Proceedings of the National Academy of Sciences, USA* **76**, 6371–5.

Pearson, C. J. & Milthorpe, F. L. (1974). Structure, carbon dioxide fixation and metabolism of stomata. *Australian Journal of Plant Physiology* **1**, 221–36.

Raschke, K. & Dittrich, P. (1977). (^{14}C) carbon-dioxide fixation by isolated leaf epidermis with stomata closed or open. *Planta* **134**, 69–75.

Rutter, J. C. (1978). An ultrastructural and biochemical study of the leaf epidermis of *Paphiopedilum* spp. with reference to stomatal functioning. M.Sc. Thesis, Stirling University, U.K.

Rutter, J. C. & Willmer, C. M. (1979). A light and electron microscopy study of the epidermis of *Paphiopedilum* spp. with emphasis on stomatal ultrastructure. *Plant, Cell and Environment* **2**, 211–19.

Schnabl, H. (1977). Isolation and identification of soluble polysaccharides in epidermal tissue of *Allium cepa*. *Planta* **135**, 307–11.

Schnabl, H. (1978). The effect of Cl⁻ upon the sensitivity of starch containing and starch-deficient stomata and guard cell protoplasts towards potassium ions, fusicoccin and abscisic acid. *Planta* **144**, 95–100.

Schnabl, H. & Ziegler, H. (1977). The mechanism of stomatal movement in *Allium cepa* L. *Planta* **136**, 37–43.

Steinberger, A. L. (1922). Regulation des osmotischen Wertes in den Schliesszellen. *Biologisches Zentralblatt* **42**, 405–19.

Thorpe, N., Brady, C. J. & Milthorpe, F. L. (1978). Stomatal metabolism: primary carboxylation and enzyme activities. *Australian Journal of Plant Physiology* **5**, 485–93.

Thorpe, N. & Milthorpe, F. L. (1977). Stomatal metabolism: CO_2 fixation and respiration. *Australian Journal of Plant Physiology* **4**, 611–21.

Thorpe, N., Willmer, C. M. & Milthorpe, F. L. (1979). Stomatal metabolism: carbon dioxide fixation and labelling patterns during stomatal movements in *Commelina cyanea*. *Australian Journal of Plant Physiology* **6**, 409–16.

Travis, A. J. & Mansfield, T. A. (1977). Studies of malate formation in 'isolated' guard cells. *New Phytologist* **78**, 541–6.

Van Kirk, C. A. & Raschke, K. (1978a). Release of malate from epidermal strips during stomatal closure. *Plant Physiology* **61**, 474–5.

Van Kirk, C. A. & Raschke, K. (1978b). Presence of chloride reduces malate production in epidermis during stomatal opening. *Plant Physiology* **61**, 361–4.

Willmer, C. M. & Dittrich, P. (1974). Carbon dioxide fixation by epidermal and mesophyll tissues of *Tulipa* and *Commelina*. *Planta* **117**, 123–32.

Willmer, C. M. & Firth, P. (1980). Carbon isotope discrimination of epidermal tissue from leaves of various plants. *Journal of Experimental Botany* **31**, 1–5.

Willmer, C. M., Pallas, J. E. Jr. & Black, C. C. (1973*a*). Carbon dioxide metabolism in leaf epidermal tissue. *Plant Physiology* **53**, 448–52.

Willmer, C. M., Kanai, R., Pallas, J. E. Jr. & Black, C. C. (1973*b*). Detection of high levels of phosphoenolpyruvate carboxylase in leaf epidermal tissue and its possible significance in stomatal movements. *Life Science* **12**, 151–5.

Willmer, C. M. & Rutter, J. C. (1977). Malic acid metabolism in guard cells. *Nature* **269**, 327–8.

Willmer, C. M., Thorpe, N., Rutter, J. C. & Milthorpe, F. L. (1978). Carbon dioxide fixation in attached and detached epidermis of *Commelina*. *Australian Journal of Plant Physiology* **5**, 767–78.

E. ZEIGER

Novel approaches to the biology of stomatal guard cells: protoplast and fluorescence studies

Introduction

Stomatal biology, a classical subject of the botanical sciences, has been studied at various levels of organization, from complex, integral approaches at the whole-plant level (Cowan & Farquhar, 1977) to investigations of enzymatic activity at the molecular level (Outlaw *et al*., 1980). In this paper we examine stomatal biology at the cellular level with emphasis on the organization and physical properties of the guard cell. Studies at this level are important, since stomatal movement, a phenomenon itself at another level of organization as it includes the extracellular pore and the water relationships of the whole epidermal tissue, depends primarily on the properties of the guard cells.

We have developed two new approaches to the study of the guard cell, the isolation of protoplasts and studies of their intrinsic fluorescence, using methods that overcome some of the usual difficulties of working with guard cells in isolation and which provide new insights into their physiological properties. I shall review the methodology of these studies and present some of the more significant results. Also, I shall discuss some of these findings in the framework of a chemiosmotic mechanism of energy transduction within the guard cells.

Isolation of guard cell protoplasts

The development of convenient methods for the isolation of plant cell protoplasts provides a new tool for probing the physiology and development of guard cells. We have devised a procedure for isolating guard cell protoplasts from leaf tissue by manipulating digestion times and washing procedures, which takes advantage of the exceptionally thick cell wall of the guard cell (Zeiger & Hepler, 1976).

Guard cell protoplasts have been prepared both in microchambers and in test tubes. The microchamber technique (Zeiger & Hepler, 1976; see Fig. 5) yields a small number of protoplasts (from ten or twenty to several hundred) but has distinct advantages. It minimizes manipulation because the enzymatic digestion, washing and any change of solutions required for experimental treatments are accomplished by perfusing the chambers through their O-ring with hypodermic needles and small perfusion pumps. We use a low-cost perfusion pump (Razel Scientific, Stamford, Connecticut, USA) that offers a wide range of perfusion rates. With an effective volume of the microchamber of $0.65 \, cm^3$, solutions within the chamber can be replaced in very short times. Two pumps connected in parallel with a valve allow sequential treatments with different solutions without interrupting the perfusion.

The chambers also allow continuous optical observation, making it possible to watch the response of the protoplasts to the digestion procedure or to any experimental treatment. For example, we were able to quantify the hydrolysis of the cellulose in the guard cells by measuring the loss of birefringence of their walls with polarized light at different digestion times (Zeiger & Hepler, 1976).

An epidermal peel or paradermal slice is mounted on one coverslip of the microchamber, with the cuticle toward the coverslip. Paradermal slices are used when the species under study does not readily yield epidermal peels or strips, or to maximize the yield of protoplasts since many guard cells are irreversibly damaged during peeling.

A cellulolytic enzyme in the appropriate osmoticum is then introduced into the chamber. Intact guard cells do not withstand the abrupt changes in osmotic conditions that result when they are exposed to the strong osmoticum (c. 0.7 M) usually used to prepare mesophyll protoplasts. If these concentrated osmotica are used at the beginning of the digestion procedure, severe damage to the guard cell plasmalemma takes place (Zeiger & Hepler, 1976). This difficulty is obviated by starting the digestion in relatively dilute osmoticum, usually 0.2–0.3 M mannitol and then raising it to the required osmotic strength. With young guard cells of onion, 0.23 M mannitol plus the osmotic contribution of a 4% (w/v) cellulytic enzyme will suffice, whereas guard cells from more mature leaves may require more concentrated osmotica. Cellulysin (CalBiochem, La Jolla, California) has proved to be an effective digestion enzyme for protoplast preparation from all species studied to date. Driselase hydrolyzes cellulose much faster, but is quite toxic (Zeiger & Hepler, 1976).

Three classes of protoplasts are obtained, derived from mesophyll, epidermal and guard cells. Mesophyll protoplasts are the first to appear, whereas guard cells, because of their remarkable, thickened walls, release their protoplasts last. These different digestion times make it possible to separate the three populations by washing at appropriate times (Zeiger & Hepler, 1976). At room temperature, guard cell protoplasts from young leaves are released in an overnight digestion; the digestion time can be shortened substantially by incubating at 25 or 30 °C.

The three classes of protoplasts are clearly distinguishable by morphological criteria. Mesophyll protoplasts are conspicuous because of the large number of chloroplasts. In onion, epidermal protoplasts are highly vacuolated and average 45 μm in diameter, whereas the guard cell protoplasts are 18 to 25 μm in diameter and have dense, granular cytoplasm (Fig. 1a).

The released guard cell protoplasts are washed and stabilized in 0.5–0.7 M mannitol with 0.5 mM $CaCl_2$. They usually remain adhering to the undigested cuticle, in the vicinity of the stomatal pore. We had once reasoned that some undigested portion of the cell matrix kept the protoplasts adhering to the cuticle (Zeiger & Hepler, 1976). Later studies, however, showed that protoplasts far enough from the pore to be outside the area once occupied by the walls withstood flow rates of 1.5 cm^3 min^{-1} without becoming detached. Furthermore, isolated vacuoles released from protoplasts, as discussed below, also adhere to the cuticle. It is possible, therefore, that electrostatic charges cause the protoplasts or their organelles to adhere to the cuticle.

The other method of preparing guard cell protoplasts involves the use of test tube and centrifugation (Schnabl, Bornman & Ziegler, 1978). This procedure generates a large number of protoplasts and it is the method of choice for biochemical analysis because of the higher yields. In order to separate the guard cell protoplasts from the protoplasts from other cells, the preparations are centrifuged early in the digestion procedure, when most of the mesophyll and epidermal cell protoplasts have been released, but the guard cells are still intact. For a high degree of purity, further separation in some type of gradient system is required (Boller & Kende, 1979).

Guard cell protoplasts have been isolated from *Allium cepa* L. (Schnabl *et al.*, 1978; Zeiger & Hepler, 1976); *Vicia faba* L. (Schnabl *et al.*, 1978); *Nicotiana tabacum* L. (Zeiger & Hepler, 1976); and *Paphiopedilum harrisianum* Hort. (Fig. 1c). The methods described should be applicable to many other species with only minor modifications. In our hands, younger leaves usually give better results; there are indications that the nutritional status of the plants might influence the success of protoplast release and their stability (Schnabl, 1978 and personal communication). It is noteworthy that we failed to obtain protoplasts from barley leaves because the digestion would hydrolyze both ends of the dumbbell-shaped guard cells but not the central thickened portion. The enlarged vacuoles located at both ends of the cell remain jacketed by the undigested portion of the wall and prevent protoplast release (E. Zeiger, unpublished).

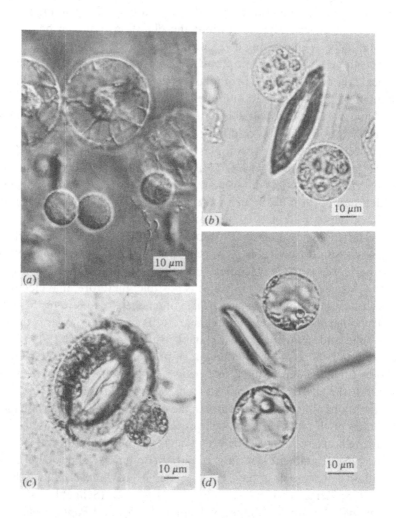

Fig. 1. (a). Epidermal and guard cell protoplasts from onion. Epidermal cell protoplasts are highly vacuolated and average 45 μm in diameter; guard cell protoplasts have a granular, dense protoplasm and average about 20 μm in diameter. (b). Guard cell protoplasts of *Vicia faba* in the vicinity of the undigested stomatal pore. The large intracellular organelles are chloroplasts. (c). Guard cells of *Paphiopedilum harrisianum*, one intact and one with its protoplast released. (d). Guard cell protoplasts from *Nicotiana tabacum*.

Experiments with guard cell protoplasts

Protoplast swelling

Since they lack a cell wall, protoplasts respond to osmotic changes in the medium with fluctuations in volume. If the osmoticum is largely impermeable, like mannitol, only water is exchanged whereas salts used as osmotica are probably taken up as well. Since guard cells modulate stomatal opening and closing by changes in osmotic potential (Hsiao, 1976), one might expect that their protoplasts would show substantial changes in volume in response to conditions affecting stomatal movements. This, indeed, has been observed.

The simplest way to quantify changes in protoplast volume is to measure their diameters at different time intervals. We used a Vikers split image eye piece (Zeiger & Hepler, 1977), but a conventional micrometer should suffice. Using microchambers, the same population of protoplasts can be measured several times, thus reducing the population size required for statistical analysis.

The swelling of guard cell protoplasts in response to light has been observed in onion (Zeiger & Hepler, 1977). This swelling is KCl-dependent and averages a 50% increase in volume after 1 h illumination with white light at 50 J m^{-2} s^{-1}. Assuming that all the swelling is due to KCl uptake, we calculated that the increase in volume corresponds to a 0.2 M increase in the intracellular solute concentration, which is approximately 0.5MPa. Epidermal cell protoplasts, treated in identical conditions, do not swell (Zeiger & Hepler, 1977), indicating that the guard cell protoplasts have a specific response to light.

Since guard cells take up K^{+} and increase their volume when stomata open (Raschke, 1975), it seems likely that the light-induced swelling of their protoplasts reflects a normal physiological property of the guard cells. That interpretation is strengthened by studies with *Vicia faba* showing a very good correlation between stomatal opening and protoplast swelling induced by fusicoccin in the one hand and closing of stomata and protoplast shrinking caused by abscisic acid on the other (Schnabl *et al.*, 1978).

The spectral sensitivity of the swelling response in onion points to a specific photobiological property of the guard cells. When broad-band blue, green and red illuminations were tested for their ability to induce swelling, only blue light was effective (Zeiger & Hepler, 1977). Because of the lack of sensitivity to red light, it seems unlikely that the guard cell chloroplasts are involved in this photoresponse. Instead, the blue-light-induced swelling seems to be related to the enhanced stomatal opening caused by blue light

(Zeiger & Hepler, 1977; Zeiger, 1980). If that is the case, isolated guard cell protoplasts from onion provide a convenient system for studying the blue-light response separately from the light responses mediated by chlorophyll.

How can blue light cause protoplast swelling? We proposed that light drives proton extrusion at the plasmalemma, thus creating electrical and pH gradients across the membrane. These gradients can then be used to drive the uphill transport of K^+ and Cl^-. The ion uptake will in turn increase the osmotic potential of the protoplast, cause water uptake and lead to swelling (Zeiger & Hepler, 1977). In fact, this chemiosmotic mechanism seems the most adequate hypothesis to explain active ion transport during stomatal opening, and many of the known physiological properties of the guard cells are consistent with its operation (Zeiger, Moody, Hepler & Varela, 1977; Zeiger, Bloom & Hepler, 1978).

Organelle isolation

The ability to obtain guard cell protoplasts also allows us to isolate guard cell organelles. This is important because many of the properties of the guard cells probably depend on special characteristics of their organelles, and their study in isolation has been very difficult in the past.

Fig. 2. An isolated vacuole from a guard cell of *Vicia faba* kept in a microchamber. The remains of the lysed protoplast can also be seen.

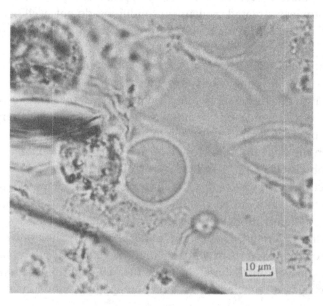

We first reported on some properties of isolated guard cell vacuoles in a study using microchambers (Zeiger & Hepler, 1977). We found that when protoplasts burst, intact vacuoles remain stable for some time, indicating that the tonoplast is more resistant to lysis than is the plasmalemma. Thus, guard cell vacuoles can be obtained by lowering the osmolality of the bathing medium to between 0.1 and 0.2 M mannitol for a few minutes until the protoplasts burst and then raising it again to 0.5 M mannitol after vacuolar release. If the lowering of the osmolality is abrupt, the vacuoles frequently fragment, forming smaller, yet apparently closed, vesicles. Slow perfusion rates (lower than $1 \, cm^3 \, min^{-1}$) usually yield large, single vacuoles (Fig. 2).

As discussed in the following section, we were able to localize a newly discovered green, intrinsic fluorescence of the guard cells (Zeiger & Hepler, 1977, 1979) using isolated vacuoles. The same method should also be useful for measuring osmotic and electrical properties of the tonoplast and the prevailing intravacuolar pH.

Finally, protoplast preparations should also be the method of choice to isolate highly purified guard cell chloroplasts and mitochondria.

Guard cell differentiation

Guard cell protoplasts cultured in a simple medium remain alive for several days (Zeiger & Hepler, 1976), indicating that in-vitro culture of guard cell protoplasts could be successful. The extreme specialization of the guard cells raises questions regarding their ability to divide, dedifferentiate and redifferentiate that should make further studies with guard cell protoplasts most worthwhile.

The fluorescence of guard cells

Fluorescence studies are an important tool for photochemical and photobiological investigations. A chemical species excited by light can re-emit energy at a longer wavelength, and the absorption and emission characteristics of the resulting fluorescence are useful parameters for the identification of the involved species. Of special interest to biologists are the cases in which electron transfers are implicated. In those instances, fluorescence usually competes with photochemistry, so that, when photochemical processes are inhibited, fluorescence intensity increases, providing a powerful experimental means for dissecting the pathways of energy transfer under examination. In this section I shall discuss some of the fluorescing properties of the guard cells, the red fluorescence of their chloroplasts and a recently discovered green vacuolar fluorescence.

Fluorescence of guard cell chloroplasts

Chloroplasts are a remarkably constant feature of guard cells (Meidner & Mansfield, 1968), indicating that they have a central function in stomatal physiology. On the other hand, our knowledge of the properties of guard cell chloroplasts is still limited and often contradictory, especially with regard to their ability to fix carbon dioxide. Some of the available data indicate an absence of Calvin cycle intermediates or the corresponding enzymes (Raschke & Dittrich, 1977; Outlaw et al., 1980; H. Schnabl, personal communication), whereas evidence for photosystem II activity based on oxygen evolution measurements has been found in Vicia faba. In our recent work on the fluorescence of guard cell chloroplasts we have obtained some new results that pertain to this matter.

We have found high levels of mesophyll chloroplast contamination in epidermal peels observed under fluorescence microscopy. Epidermal peels from Vicia faba, Allium cepa and Tulipa sp. all showed abundant red fluorescing organelles that are most likely to be chloroplasts, outside the guard cells proper (Fig. 3). These chloroplasts were not within intact mesophyll cells and were difficult or impossible to observe under bright-field microscopy (Fig. 3a) (Zeiger, Armond & Mellis, 1980). This contamination complicates the interpretation of experiments with epidermal peels, generally assumed to be free of chloroplasts outside the guard cells (Lurie, 1977; Raschke & Dittrich, 1977; Dittrich & Raschke, 1977a, b), and emphasizes the value of fluorescence microscopy as a means of evaluating contamination of epidermal peels by mesophyll chloroplasts.

We have also obtained low temperature emission spectra from epidermal peels and room temperature spectra from single guard cells. The room temperature emission spectrum from a guard cell in an epidermal peel from V. faba, obtained with a microspectrophotometer (Fig. 4a) shows a distinct peak at around 683 nm and a broad shoulder in the 720 to 750 nm region (E. Zeiger & P. Armond, unpublished). Spectra obtained at room temperature offer limited resolution but should be valuable for studies of changes in fluorescence intensity in single guard cells in the presence of inhibitors of photosynthesis (Zankel & Kok, 1972). Fig. 4 also shows the low temperature emission spectrum from an epidermal peel preparation of V. faba frozen in liquid nitrogen ($-196\,°C$), with two major peaks at around 687 and 733 nm. These two peaks are generally attributed to the activity of photosystems II and I, respectively (Satoh & Butler, 1978). However, as mentioned above, epidermal peels from V. faba are heavily contaminated with mesophyll chloroplasts, which would contribute to the fluorescence signal. In our attempt to evaluate the fluorescence, we also obtained low temperature spectra of an albino portion of Chlorophytum comosum Baker ('spider

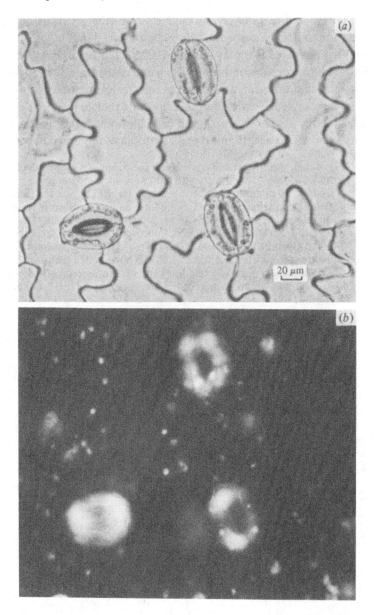

Fig. 3. Bright field (*a*) and fluorescence (*b*) micrographs of an epidermal peel of *Vicia faba*. The small dots in the fluorescence micrograph are fluorescing chloroplasts from broken mesophyll cells. These chloroplasts are very hard to detect under bright field. The fluorescence picture was taken on Kodak Ektachrome Professional colour film and then reproduced in black and white. The images of the guard cells are blurred because they are above the focal plane of the chloroplasts.

plant') leaves and of isolated chloroplasts from the green tissue. Careful examination under fluorescence microscopy showed that in the albino portion of the variegated leaf only guard cells had chloroplasts. As seen in Fig. 4b, an emission spectrum of the albino portion of the leaf shows peaks at 686 and 740 nm that are very similar to the spectra of mesophyll chloroplasts from the same leaf. The data, therefore, indicate that, at least in C. comosum and presumably also in V.faba, guard cell chloroplasts possess both photosystems I and II. Hence, if guard cell chloroplasts are indeed unable to fix

Fig. 4. (a). Interrupted line: emission spectrum of a single guard cell from *Vicia faba* excited with blue light at room temperature. Nanospec/10 microfluorospectrophotometer, BG 12 exciting filter and 450 nm dichroic beam splitter. Solid line: emission spectrum from a suspension of epidermal peels of *Vicia faba* frozen in liquid nitrogen (−196 °C). Perkin Elmer MPF-32 fluorospectrophotometer, excitation at 440 nm; Corning filters 4–96 and 2–58. (b). Low temperature (−196 °C) emission spectra of a slice from an albino portion of a leaf from *Chlorophytum comosum* (solid line) and of a suspension of chloroplasts isolated from a green portion of the leaf (interrupted line). Conditions as in Fig. 4a, solid line.

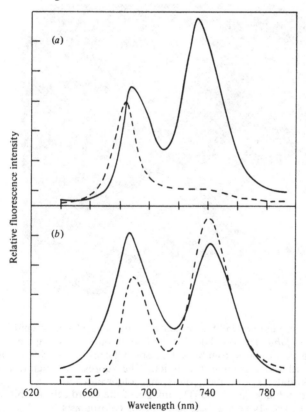

carbon dioxide, that deficiency cannot be explained by a lack of photosystem II activity. What, then, would be the function of chloroplasts in the guard cells? We have suggested that an important role would be to provide ATP and/or reducing power to drive proton extrusion and sustain active ion transport during stomatal opening (Zeiger *et al.*, 1977, 1978). If this were the case, guard cell chloroplasts might be the main source of energy for stomatal opening in the light.

Green fluorescence of guard cell vacuoles

When excited with blue light, the vacuoles of guard cells of some species fluoresce in the green. This fluorescence was originally discovered in *A. cepa* (Zeiger & Hepler, 1977, 1979) and later found in *V. faba* (Zeiger, 1980). Guard cell protoplasts from *Paphiopedilum harrisianum* also exhibit a green fluorescence, but its cellular localization remains uncertain (Zeiger, unpublished).

The green fluorescence can be conveniently studied under fluorescence microscopy, using a vertical illuminator for incident fluorescence, which provides intense illumination under high optical magnification (Zeiger & Hepler, 1979). Fluorescence intensity measurements and emission spectra were obtained with a Nanometrics single-cell microspectrophotometer (Sunnyvale, California, USA). Guard cells from epidermal peels, their protoplasts or their isolated vacuoles were observed in microchambers under perfusion, as described in a preceding section. The perfusion chambers are particularly useful for fluorescence studies because they allow rapid changes in the bathing medium while the cells are under continuous observation and also eliminate the spurious fluorescence sometimes observed when using a conventional microscope slide and coverglass. A diagram of this experimental system is shown in Fig. 5.

Onion guard cells show the green vacuolar fluorescence when bathed in water. Isolated vacuoles, obtained as described before, retain the fluorescence, and indirect evidence suggests that the fluorescing compound is located in the tonoplast (Zeiger, 1980 and unpublished). Emission and excitation characteristics are indicative of a flavin (Zeiger, 1980).

Studies showing that guard cells from *Paphiopedilum* spp. are functional in spite of the lack of chloroplasts (Nelson & Mayo, 1975; Rutter & Willmer, 1979) have generated considerable interest in this genus. Our recent studies with guard cells from *P. harrisianum* are consistent with these reports. Single guard cells observed at room temperature under incident fluorescence microscopy lack the characteristic red fluorescence of chloroplasts. Furthermore, epidermal peels free from mesophyll chloroplasts and frozen in liquid nitrogen were observed in a fluorospectrophotometer, and they failed to show either the 687 or the 735 nm peaks. Hence those guard cells appear to lack both photosystems I and II (Zeiger *et* al., 1980).

On the other hand, guard cells in microchambers observed under fluorescence microscopy show a distinct, abundant green fluorescence over most of the guard cells. Its characteristics are, however, different from the one seen in onion, where the absence of fluorescence over the nuclei indicated that the fluorescence is localized in the vacuole (Zeiger & Hepler, 1979). The green fluorescence in *P. harrisianum* was further investigated by isolating guard cell protoplasts. Upon protoplast release, most of the fluorescence remained associated with the cell walls, although the protoplasts retained a faint, yet unequivocal, green fluorescence (E. Zeiger, unpublished). Further

Fig. 5. Diagram of the experimental set-up used to study the fluorescence of stomatal guard cells. Epidermal peels are mounted in microchambers and perfused with the perfusion pump. A variable slit controls the area of the cell exposed to the spectrophotometer. The optical system is equipped with incident fluorescence and a 50 W mercury lamp. A Nanospec/10 microfluorospectophotometer with a gallium-arsenide photomultiplier and a motor-driven diffraction grating monochromator and a chart recorder are used for quantitative measurements.

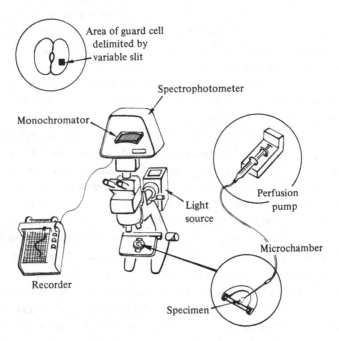

studies will possibly determine the functional role of this fluorescence and its relationship to the green fluorescence of guard cells of *V. faba* and onion.

Guard cells from *V.faba* lack the green vacuolar fluorescence when bathed in water but exhibit it in the presence of ammonia, CCCP and nigericine (Zeiger, 1980). The spectral characteristics and the cellular localization of the green fluorescence are similar in onion and *V. faba,* indicating that the fluorescing compounds might be identical. On the other hand, the fact that onion guard cells fluoresce in an aqueous solution, whereas guard cells of *V. faba* require an inductive treatment, indicates that some of the physiological properties affecting the fluorescence must be different. The nature of the treatments capable of inducing the fluorescence in *V. faba* suggests that one basic difference is the prevailing intracellular pH, the one in onion presumably being more alkaline (Zeiger, 1980).

Recent findings showing that fusicoccin can also induce the vacuolar fluorescence in *V. faba* (E. Zeiger, unpublished) are consistent with the suggestion that the vacuolar fluorescence is pH-dependent. Guard cells in epidermal peels of *V. faba* mounted in microchambers were perfused with water at 1.5 cm^3 min^{-1} and then exposed to 5.0 μM fusicoccin in 20 mM K$_2$SO$_4$. The green vacuolar fluorescence was detected within a few minutes and could not be induced by either fusicoccin or K$_2$SO$_4$ alone.

Fusicoccin is known to enhance proton extrusion in many systems, including guard cells (Marrè, 1979). These proton fluxes hyperpolarize the cells and presumably also cause intracellular alkalinization. In our experimental conditions this alkalinization should be further enhanced by the absence of chloride ions in the medium, which would preclude a Cl$^-$/OH$^-$ exchange (Zeiger *et al.,* 1978). The induction of the fluorescence by fusicoccin is therefore consistent with an intracellular alkalinization, although the possible effects of electrical changes should also be considered.

These studies on the vacuolar fluorescence point to its potential as a probe of the physiological properties of the guard cells. Further experimental work should benefit from its use as a sensitive intracellular pH indicator, as a possible means of characterizing the blue light response of stomata and as a potential tonoplast marker.

In conclusion, protoplast and fluorescence studies demonstrate the advantages of a cellular approach in the quest for an understanding of the contribution of guard cells to stomatal opening and closing.

I thank Eleanor Crump for editing the manuscript and Professor E. Marrè for a generous gift of fusicoccin. This work was supported by NSF Grant PCM-77-17642.

References

Boller, T. & Kende, H. (1979). Hydrolytic enzymes in the central vacuole of plant cells. *Plant Physiology* **63**, 1123–32.

Cowan, I. R. & Farquhar, G. D. (1977). Stomatal function in relation to leaf metabolism and environment. *31st Symposium of the Society for Experimental Biology,* pp. 471–505. Cambridge University Press.

Dittrich, P. & Raschke, K. (1977a). Malate metabolism in isolated epidermis of *Commelina communis* L. in relation to stomatal functioning. *Planta* **134**, 77–81.

Dittrich, P. & Raschke, K. (1977b). Uptake and metabolism of carbohydrates by epidermal tissue. *Planta* **134**, 83–90.

Hsiao, T. C. (1976). Stomatal ion transport. In *Encyclopedia of Plant Physiology,* New Series, ed. U. Luttge & M. G. Pitman, Vol. 2, part B, pp. 195–221. Berlin-Heidelberg-New York: Springer-Verlag.

Lurie, S. (1977). Photochemical properties of guard cell chloroplasts. *Plant Science Letters* **10**, 219–23.

Marrè, E. (1979). Fusicoccin: a tool in plant physiology. *Annual Review of Plant Physiology* **30**, 273–88.

Meidner, H. & Mansfield, T. A. (1968). *Physiology of Stomata.* London: McGraw Hill.

Nelson, S. D. & Mayo, J. M. (1975). The occurrence of functional non-chlorophyllous guard cells in *Paphiopedilum* spp. *Canadian Journal of Botany* **53**, 1–7.

Outlaw, W. H., Manchester, J., Di Camelli, C. A., Randall, D. D., Rapp, B. & Veith G. M. (1979). Photosynthetic carbon reduction pathway is absent in chloroplasts of *Vicia faba* guard cells. *Proceedings of the National Academy of Science, USA* **76**, 6371–75.

Raschke, K. (1975). Stomatal action. *Annual Review of Plant Physiology* **26**, 309–40.

Raschke, K. & Dittrich, P. (1977). [^{14}C] carbon dioxide fixation by isolated epidermis with stomata closed or open. *Planta* **134**, 69–75.

Rutter, J. C. & Willmer, C. M. (1979). A light and electron microscopy study of the epidermis of *Paphiopedilum* spp., with emphasis on stomatal ultrastructure. *Plant, Cell and Environment* **2**, 211–19.

Satoh, K. & Butler, W. L. (1978). Competition between the 735 nm fluorescence and the photochemistry of photosystem I in chloroplasts at low temperature. *Biochimica et Biophysica Acta* **502**, 103–10.

Schnabl, H. (1978). The effect of Cl$^-$ upon the sensitivity of starch-containing and starch-deficient stomata and guard cell protoplasts towards potassium ions, fusicoccin and abscisic acid. *Planta* **144**, 95–100.

Schnabl, H., Bornman, Ch. & Ziegler, H. (1978). Studies on isolated starch-containing (*Vicia faba*) and starch-deficient *(Allium cepa)* guard cell protoplasts. *Planta* **143**, 33–9.

Zankel, K. L. & Kok, B. (1972). Estimation of pool sizes and kinetic constants. *Methods in Enzymology* **24**, part B, 218–38.

Zeiger, E. (1980). The blue light response of stomata and the green vacuolar fluorescence of guard cells. In *The Blue Light Syndrome,* ed. H. Senger, pp. 629–36. Heidelberg: Springer-Verlag.

Zeiger, E., Armond, P. & Melis, A. (1980). Fluorescence properties of guard cell chloroplasts; evidence for linear electron transport and light harvesting pigments of PS I and PS II. *Plant Physiology* **66** (in press).

Zeiger, E., Bloom, A. & Hepler, P. K. (1978). Ion transport in stomatal guard cells: a chemiosmotic hypothesis. *What's New in Plant Physiology* **9**, 29–32.

Zeiger, E. & Hepler, P. K. (1976). Production of guard cell protoplasts from onion and tobacco. *Plant Physiology* **58**, 492–8.

Zeiger, E. & Hepler, P. K. (1977). Light and stomatal function: blue light stimulates swelling of guard cell protoplasts. *Science* **196**, 887–9.

Zeiger, E. & Hepler, P. K. (1979). Blue light-induced, intrinsic vacuolar fluorescence in onion guard cells. *Journal of Cell Science* **37**, 1–10.

Zeiger, E., Moody, W., Hepler, P. & Varela, F. (1977). Light sensitive membrane potentials in onion guard cells. *Nature* **270**, 270–1.

T. A. MANSFIELD, A. J. TRAVIS AND R. G. JARVIS

Responses to light and carbon dioxide

Conceptual separation of effects of light and carbon dioxide

The responses of stomata to light and carbon dioxide are conveniently considered together, because both appear to make a substantial contribution to the daily cycle of opening and closing. Although, in analyses of the opening process, the relative sizes of the contributions of light and reduced carbon dioxide concentration have been difficult to determine, there is now general agreement that there is a clear distinction between them. Many experimental results have suggested that we must recognize a light-independent carbon dioxide response, and a carbon dioxide-independent light response. The evidence for the former is simple and indisputable: stomata open in darkness in response to the removal of carbon dioxide from the leaf's internal atmosphere. This has been demonstrated for many species, and although the opening achieved is sometimes small it can be enhanced by a temperature rise coinciding with the removal of carbon dioxide. Indeed this combination of treatments can produce opening which is only a little less than that in light (Fig. 1).

There is reason to believe that this ability to respond to the removal of carbon dioxide in the absence of light is the cause of the night opening of stomata in succulents (Neales, 1970). In these plants a massive fixation of carbon dioxide occurs at night leading to a reduction in intercellular carbon dioxide concentration comparable to that in other plants in light. The succulents are an exception to the general rule of stomatal opening by day and closing by night, but they provide confirmation that light is not necessary for stomatal carbon dioxide responses.

The evidence that light can induce opening independently of changes in carbon dioxide concentration comes mainly from studies of the extent of opening in different wavelengths of light. Kuiper's (1964) action spectrum for opening on epidermal strips of *Senecio odoris* Defl. showed a marked peak in the blue region, substantially higher than that in the red. Mansfield &

Meidner (1966) found that when leaves of *Xanthium strumarium* L. were maintained in a closed air circuit of the kind used for determining carbon dioxide compensation, blue wavelengths could stimulate stomatal opening even when the irradiance was below the light compensation point for photosynthesis (Fig. 2a). At the lowest irradiance, carbon dioxide compensation was not achieved and the closed circuit had to be ventilated to prevent carbon dioxide accumulating. This experiment showed that opening could be induced by blue light despite the presence of a high carbon dioxide concentration in the intercellular spaces, and it can be inferred that in natural conditions light can, by virtue of the effects of its blue component, cause stomatal opening to begin at low levels of irradiance that are insufficient to produce a net photosynthetic intake of carbon dioxide. In Fig. 2b some of the tabulated results of Mansfield & Meidner (1966) have been plotted graphically to reveal the contrast between the relative effectiveness of red and blue light in causing stomatal opening (cf. Fig. 2b) and in achieving a low carbon dioxide concentration in the intercellular space, as indicated by the level of carbon dioxide compensation. If light were producing opening purely by causing a photosynthetic consumption of carbon dioxide in the leaf, the ability of a light source to induce opening should be clearly related to the amount of photosynthesis it produces. This is manifestly not the case, both from observations like those illustrated in Fig. 2, and from the action spectra subsequently obtained by several workers confirming the findings of Kuiper (1964). The fact that stomata respond to

Fig. 1. Continuous porometer trace (solid line) showing that the stomata of *Xanthium strumarium* can open widely in darkness if the temperature is increased and the leaf's internal atmosphere ventilated with carbon dioxide-free air. The temperature rise was from 27 to 36 °C; opening of control stomata at 27 °C was determined at 30 min intervals (lower broken line). The mean opening at 27 °C in light (irradiance = 60 W m^{-2}) is shown by the upper broken line. After Mansfield (1976).

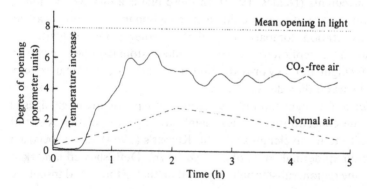

changes in carbon dioxide concentration in light does, however, indicate that photosynthesis must exert some control over opening. The action spectrum for opening on the intact leaf would be expected to be derived partly from the stimulation of photosynthesis, with approximately equal peaks in the blue and red wavelengths (Balegh & Biddulph, 1970) and from the carbon dioxide-independent action of blue light. The effect of blue light on this process closely matches that found elsewhere in green plants and fungi

Fig. 2. (*a*). Effect of irradiance of red and blue light on stomatal opening in *Xanthium strumarium*. After Mansfield & Meidner (1966). (*b*). Data of Mansfield & Meidner (1966) replotted to show changes in the level of carbon dioxide compensation brought about by the irradiance of red and blue light. Broad band sources were used with peak irradiances at 475 and 640 nm.

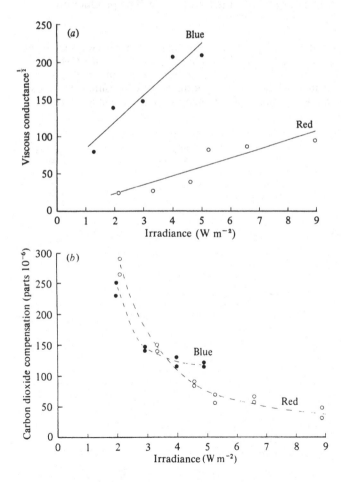

(Smith, 1975) and is probably due to the ubiquitous flavin photoreceptor that controls a wide range of physiological activities. This receptor does not function in the red region of the spectrum, and hence we might suppose that in the intact leaf, stomatal opening in red light would be solely the result of photosynthetic removal of carbon dioxide. Kuiper did, however, obtain some opening in red light using isolated epidermis of *Senecio odoris* and in our own recent research we have found that the same is true for epidermis of *Commelina communis* L. that is not contaminated with mesophyll (Fig. 3). This species is the one in which Raschke & Dittrich (1977) sought evidence of the ability of guard cells to reduce carbon dioxide photosynthetically. They found no such ability, and hence we are obliged to conclude that the stomatal opening induced by red light does not solely result from the removal of carbon dioxide by photosynthesis. It seems that the absorption of light by guard cell chloroplasts may be used to provide some of the energy consumed by the processes responsible for opening. The production of ATP

Fig. 3. Stomatal opening in red or blue light. Epidermis was incubated for 3 h at 25 °C on 10 mol m^{-3} MES buffer (pH 6.15 with KOH) containing 50 mol m^{-3} KC1. Each point is the mean of 3 replicate experiments in which 30 stomatal apertures were measured. The bars represent ± s.e. for three measurements of mean stomatal aperture.

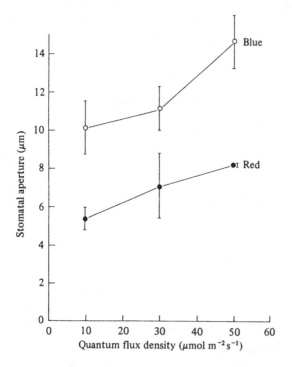

by cyclic photophosphorylation in guard cells and its involvement in the opening process is clearly implied by experiments with metabolic inhibitors (Humble & Hsiao, 1970; Willmer & Mansfield, 1970; Raghavendra & Das, 1972). By contrast, inhibitors of non-cyclic electron flow have not been found to be effective in experiments where care has been taken to ensure that secondary effects on the stomata, resulting from accumulation of carbon dioxide consequent upon the inhibition of photosynthesis, did not occur (Allaway & Mansfield, 1967; Willmer & Mansfield, 1970; Humble & Hsiao, 1970).

It is not, however, possible to accept the view that light has a *primary* role as an energy source for stomatal opening. There is much to suggest that this is not the case. In particular, we can draw attention to the wide opening in response to carbon dioxide removal in the dark (Fig. 1) and in blue light at low levels of irradiance (Fig. 2). The recent studies of lady's slipper orchids (*Paphiopedilum spp.*), which possess functional stomata but whose guard cells do not contain chloroplasts (Nelson & Mayo, 1975; Rutter & Willmer, 1979) confirm that opening in the light is not dependent on the provision of energy by chloroplasts. The ratio of mitochondria to chloroplasts is four times higher in guard cells than in mesophyll cells (Allaway & Milthorpe, 1976) and this fact combines with the accumulated physiological evidence to suggest that they are the main, and the standby, generators of the ATP needed to power the changes in turgor of guard cells. Thus in recognizing a role for the chloroplasts (where they are present) we must be careful not to imply that it is an essential one.

Thus an analysis of the factors operating to produce stomatal opening in light leads us to recognize a number of separate contributions.

    (1) Removal of carbon dioxide from the leaf's internal atmosphere by photosynthesis in the mesophyll.

    (2) An action of blue light which is independent of (1) and which may not necessarily involve chloroplasts.

    (3) An effect of light which may depend upon the ATP generating ability of guard cell chloroplasts.

(1) and (3) would be expected to combine to produce an action spectrum for stomatal opening in the intact leaf similar to that for photosynthesis, upon which (2) would be superimposed to produce a higher peak in the blue region than in the red. On isolated epidermis (1) will be absent, but (3) will combine with (2) to produce an action spectrum for opening similar to that of the whole leaf. Since (3) probably involves only cyclic photophosphorylation, wavelengths in the red/far-red region that do not permit photosynthetic carbon dioxide fixation may be expected to be effective in producing stomatal opening.

Mechanisms behind the responses

How do these different contributions combine to produce opening in light? This question cannot be answered now with much more precision than in 1954 when Heath and Russell attempted to analyse the opening processes and separate out the various controlling factors. The quarter of a century that has passed has brought great advances in our understanding of the mechanisms behind the turgor changes in guard cells, but we are still very uncertain about the relationship between the agents that we have recognized as being individually capable of inducing the changes to occur. The carbon dioxide response has received most attention over the past decade, especially in relation to the control by stomata of the supply of carbon dioxide for photosynthesis. There is evidence that the magnitude of the closing response to carbon dioxide varies among species, and, within an individual plant, according to its previous history. The response becomes greater after mild water or cold stress has been experienced, and plants of some species grown carefully to avoid all such stresses are found to be almost devoid of a stomatal response to carbon dioxide (Raschke, 1979). The stomata on plants that have received different pretreatments retain the common feature of a diurnal cycle of opening in light and closing in darkness, and it can therefore be implied that the relationship between control by carbon dioxide and by other factors is variable. Raschke looks upon this variability as providing the plant with 'an opportunity to weigh the importance of signals in the attempt to reconcile the opposing priorities of CO_2 supply and water husbandry'. The recognition of such variability in the relative contributions of controlling factors draws our attention to the level at which regulation of the turgor of guard cells occurs. As far as we are aware, both light and the removal of carbon dioxide cause stomata to open by switching on the same process in the guard cells, namely an accumulation of potassium ions. The way in which ion accumulation is initiated by each factor may be different, but since there are unlikely to be alternative *mechanisms* of potassium ion uptake we are probably justified in assuming that both factors ultimately control one process. Therefore we can suggest that there may be circumstances in which potassium ion uptake into guard cells is switched on by light *or* by carbon dioxide removal, and when it has been initiated by one the action of the other becomes unnecessary. This could help us to reconcile some otherwise contrary aspects of the effects of the two factors. For example, it is possible to find circumstances in which wide opening is achieved by removal of carbon dioxide, suggesting very little necessity for a light response (Fig. 1) but, on the other hand, in an experiment of a different sort on the same species, the blue light effect can apparently override the closure that one would expect to find in high carbon dioxide (Fig. 2).

The role of potassium

In order to obtain a better understanding of the level at which light and carbon dioxide exert control, we undertook a series of experiments using detached epidermis of *C. communis*, in which the availability of potassium ions in the surrounding medium was varied while the magnitude of stomatal responses to light and carbon dioxide was determined.

Strips of epidermis from plants that had been carefully grown to minimize environmental stress (i.e. there was presumed to be a low endogenous level of ABA) were incubated in darkness or light, with air free of carbon dioxide or containing 350 parts 10^{-6} CO_2 bubbling through the medium. MES (morpholino ethane sulphonic acid) buffer containing 5 mol m^{-3} K$^+$ was used to maintain a pH of 6.15, and different amounts of potassium chloride were added in concentrations of 0 to 200 mol m^{-3}. After incubation for 3 h, stomatal apertures and malate content of the epidermis were determined. The results are shown in Fig. 4, from which it will be seen that an increase in potassium chloride concentration stimulated stomatal opening irrespective of the light or carbon dioxide regime. However, when the magnitude of the

Fig. 4. The effect of KCl concentration on (*a*) stomatal aperture and (*b*) epidermal malate concentration. Epidermis was incubated for 3 h at 25 °C in 10 mol m^{-3} MES buffer (pH 6.15 with KOH) containing various concentrations of KCl. Each point is the mean of three replicate experiments in which the malate content of the epidermis and 30 stomatal apertures were measured. The bars indicate the least significant difference between treatments means ($P = 0.05$, D.F. = 38). Open circles, light$-CO_2$; filled circles, dark$-CO_2$; open triangles, light$+CO_2$; filled triangles, dark$+CO_2$. Data from Travis & Mansfield (1979*b*).

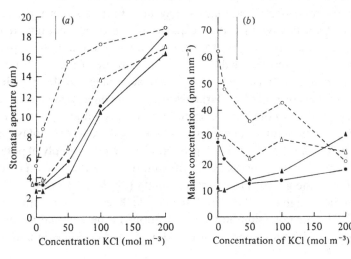

reaction of the stomata to these two factors was considered, it became clear that there was an optimum potassium chloride concentration. Fig. 5 shows the changes in stomatal aperture in response to light in the presence or absence of carbon dioxide, and of the response to carbon dioxide in the presence or absence of light. In both cases the upper curve, which is the response to each factor when the other is held at a favourable level for opening (i.e. the light response when carbon dioxide is absent, or the carbon dioxide response when light is present) shows a clear maximum at around 50 mol m^{-3} KCl.

Van Kirk & Raschke (1978) discovered that malate formation in epidermis is dependent on the amount of chloride available in the medium, and our observations confirm this, at least for the treatment most conducive to opening (minus carbon dioxide, plus light). Another effect upon malate formation that was revealed was, however, one of light and carbon dioxide which remained nearly constant throughout the range of KCl concentrations employed (Fig. 6).

The experimental results in Figs. 4 to 6 enable us to reach some conclusions about the way in which guard cell turgor is affected by light and carbon dioxide. We suggest that at potassium chloride concentrations below the 'optimum' for permitting responses to occur, the extent of stomatal opening is limited by the availability of potassium ions. This limitation has the effect of overriding the responses to the other two variables, which we suggest act by limiting the amount of potassium ion entering the guard cells. MacRobbie (1977) put forward a point of view which challenged some of the established ideas about the control of the turgor of guard cells. She pointed out that the behaviour of the guard cells as stomata open is not unlike that of other plant cells that accumulate ions, i.e. they acquire potassium and chloride ions from outside, or generate organic anions as they take up potassium ions. The peculiarity of their behaviour is their capacity to cease their net uptake of ions, lose turgor and bring about stomatal closure. Although in the experiments we have just described we did not study the closure of previously open stomata, the results can be best interpreted according to MacRobbie's suggestions, namely by regarding darkness and carbon dioxide as agents that limit the net uptake of potassium ions into guard cells, either by inhibiting the entry of potassium ion or enhancing its leakage. An external potassium chloride concentration of 50 mol m^{-3} provides an adequate supply of potassium ions for the uptake process but is sufficiently below the potassium ion concentration in the guard cells of open stomata (Penny & Bowling, 1974, found this to be as high as 448 mol m^{-3}) to allow leakage to occur if uptake is interrupted. Higher potassium ion concentrations in the external medium clearly interfere with the action of carbon dioxide and darkness in limiting

Fig. 5. The effect of KCl concentration on changes in stomatal aperture in response to carbon dioxide (*a*) and light (*b*). Each point is the difference between two treatment means from Fig. 4 and the horizontal interrupted lines represent the least significant difference between treatments means ($P = 0.05$, D.F. $= 38$). Open circles, (light $-$ CO_2) minus (light$+CO_2$); filled circles, (dark $-$ CO_2) minus (dark$+CO_2$); open triangles, (light $-$ CO_2) minus (dark $-$ CO_2); filled triangles, (light$+CO_2$) minus (dark $+$ CO_2). Data from Travis & Mansfield (1979*b*).

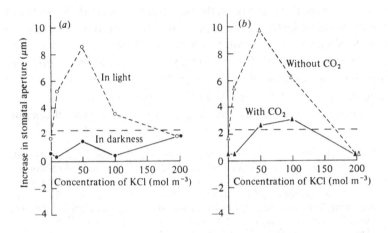

Fig. 6. The effect of KCl concentration on changes in epidermal malate content in response to carbon dioxide (*a*) and light (*b*). Each point is the difference between two treatments means from Fig. 4 and the horizontal interrupted lines represent the least significant difference between treatment means ($P = 0.05$, D.F. $= 38$). Symbols are the same as Fig. 5. Data from Travis & Mansfield (1979*b*).

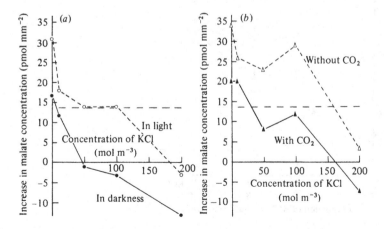

stomatal opening: they could do so either by enhancing the rate of uptake and overcoming inhibition imposed upon it by the two factors, or by reducing leakage enhanced by them. We cannot at present distinguish between these two possibilities. We can, however, suggest that the action of darkness and carbon dioxide is upon the rate of *net* accumulation of potassium ions into guard cells.

If we are correct in deducing that control is exerted specifically through potassium ion transport mechanisms, the possibility arises that the degree of control may change if guard cells are presented with alternative cations to potassium. Sodium ions are known to support stomatal opening in *C. communis* (Willmer & Mansfield, 1969) and they are apparently more effective than potassium ions in some circumstances. When we tested the ability of stomata to respond to stimuli we found, however, that there was a vast difference between these two cations. Fig. 7 shows the degree of opening

Fig. 7. The effect of different proportions of NaCl and KCl on stomatal opening in light and darkness, in the presence and absence of carbon dioxide. Epidermal strips of *Commelina communis* were incubated at 25 °C on solutions of KCl and NaCl buffered with 5 mol m^{-3} MES (pH 6.15 with Trizma base). Air (with or without CO_2) was bubbled through 10 cm^3 portions of medium at 100 cm^3 min^{-1}. Open circles, light (160 μmol m^{-2} s^{-1}) $-CO_2$; filled circles, darkness $-CO_2$; open triangles, light $+CO_2$ (350 parts 10^{-6}); filled triangles, darkness $+CO_2$. Points represent means and vertical bars 95% confidence limits based on 90 measurements (30 from each of three replicates).

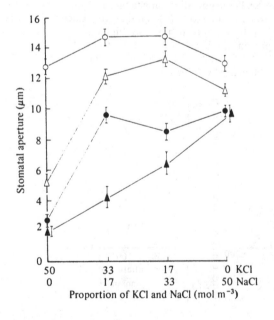

achieved on epidermis incubated in equimolar solutions of KCl and NaCl, or in mixtures of the two. It is clear that although potassium and sodium ions are equally acceptable to the guard cells as a means of increasing their turgor, the ability to respond to carbon dioxide and darkness is much reduced in NaCl. It is arguable, indeed, that the ability might be lost altogether if all the potassium were replaced by sodium ions; there must, of course, be some residual potassium remaining in the tissue and it could be upon this that the small effects of the two factors, seen in 50 mol m^{-3} NaCl, depend.

The responses of stomata to another agent known to be able to restrict their opening, abscisic acid, are similarly dependent on the potassium ion concentration available to the guard cells (Wilson, Ogunkanmi & Mansfield, 1978; Weyers & Hillman, 1979) and we have found that the response to ABA is much reduced in the presence of sodium ions (Jarvis & Mansfield, 1980). Thus it can be suggested that there is a common basis for the effect of three factors (carbon dioxide, darkness and ABA) that are naturally involved in regulating the turgor of guard cells. Their action is apparently to affect the net accumulation of potassium ions. We cannot at present say whether this is achieved by control of potassium ion uptake or its loss, or both.

A synthetic compound which binds potassium ions strongly, benzo-18-crown-6 (a cyclic polyether), has been found to be capable both of inducing stomatal closure, and of interfering with the effect of ABA (Richardson *et al.*, 1979). It seems likely that benzo-18-crown-6 may enhance leakage of potassium ion from guard cells, and the nature of its interference with the action of ABA suggests that this natural inhibitor may operate in a similar way. Perhaps further studies with benzo-18-crown-6, examining its relation to the action of carbon dioxide and to darkness as inhibitors of stomatal opening, will provide useful clues as to their mode of operation.

Our considerations here of the effects of darkness as a factor inducing stomatal closure have had the convenience of enabling us to draw comparisons with the action of two other agents, carbon dioxide and ABA. It must, however, always be borne in mind that darkness cannot be regarded as a simpler experimental treatment than light, and we have defined above (p. 123) three effects of light. In our experiments reported here on epidermal strips of *C.communis* we can disregard control of the stomata by carbon dioxide depletion as a result of photosynthetic activity, since the epidermis of this species is incapable of photosynthetic carbon dioxide reduction (Raschke & Dittrich, 1977). Two known effects of light therefore remain: the action of blue light, probably *via* a flavin photoreceptor, and a further action probably *via* cyclic electron flow and ATP production in the guard cell chloroplasts. The provision of energy by the latter would presumably affect potassium and sodium ion entry into the guard cells equally, and hence we can deduce that the closing effect of darkness, which appears to be specifically linked to potassium ion transport, is the result of a termination of that effect of light that results from absorption by the flavin photoreceptor. Thus it appears that the light response that is confined to the blue region of the spectrum, and the carbon dioxide and ABA responses, may all operate at the same level, namely by the control of potassium ion transport. Such control is likely to be resident in the guard cell membranes, perhaps in the plasmalemma adjacent to the subsidiary cells. A recognition that there is some common ground between the effects of these different agents may help us to understand interference between them. Raschke (1975a) has reported an interdependence between carbon dioxide and ABA responses, and Mansfield (1976) found that the reaction to ABA was affected by both irradiance and carbon dioxide.

Effects on anion formation in guard cells

The fact that light and carbon dioxide have more effect upon stomata in media containing KCl than in NaCl is our primary piece of evidence that these agents exert their control via a process, or processes, involved in potassium ion transport; it combines with other evidence to suggest that the main control is upon cation rather than anion movement or formation. The anions used by guard cells to balance the imported potassium ion seem to be variable in nature. In some species, such as *Zea mays* L., chloride ions play a major part (Raschke & Fellows, 1971), while in *C. communis* and *Vicia faba* L. enough malate can be produced to balance the amount of potassium ion known to enter the guard cells of open stomata (Travis & Mansfield, 1977; Van Kirk & Raschke, 1978; Raschke, 1979). When, however, there is an abundance of chloride ion available to the guard cells this may replace a substantial part of the malate (Raschke & Schnabl, 1978). If neither chloride nor malate is individually essential for the increase in guard cell turgor, it is difficult to envisage an absolute control function for either of them, of the kind that we have attributed to potassium.

There has been much discussion, both in the literature on stomata (Raschke, 1979) and elsewhere (e.g. Davies, 1979) about the source of the malate formed in guard cells and its relationship to the carbon dioxide responses of stomata. The most likely metabolic route for malate production involves the carboxylation of phosphoenolpyruvate (PEP), and some dilemma is therefore created because stomata open most widely, and hence contain most malate, in low carbon dioxide concentrations which should not favour a carboxylation reaction. The source of carbon dioxide could, of course, be an internal one, and so this question is not a major obstacle to our understanding. The question of whether PEP carboxylation can play any part in the control of stomata by carbon dioxide is a much more difficult one. Raschke (1975b, 1979) has suggested that malate formed in the cytoplasm from carbon dioxide entering the guard cells from outside could affect their turgor by causing leakage of ions. Our suggestions in the previous section do not necessarily conflict with this idea, but simply point to a similar kind of control being exerted by light and carbon dioxide. The precise stage at which they affect net ion accumulation could be different.

Travis & Mansfield (1979a) have obtained some results from experiments using the fungal toxin fusicoccin, which may offer a solution to our problems in understanding the complex effects of carbon dioxide on guard cells. Fusicoccin has long been known to induce wide stomatal opening (Graniti & Turner, 1970); it stimulates potassium ion uptake into guard cells (Squire & Mansfield, 1972) and overrides the normal closing effects of factors like darkness, carbon dioxide and ABA. It is thought to act by increasing proton extrusion from the guard cells (Raschke, 1977) and this is supported by

Fig. 8. Reversal of the CO_2 responses of stomata by fusicoccin. Epidermis was incubated for 3 h at 25 °C on 10 mol m^{-3} MES buffer (pH 6.15 with KOH). Each point is the mean of three replicate experiments in which the malate content of the epidermis (a) and 60 stomatal apertures (b) were measured. The bars represent $\pm$ s.e. for three measurements of mean stomatal aperture or malate content; standard errors smaller than the symbols are not shown. Filled circles, dark $-$ fusicoccin; open circles, light $-$ fusicoccin; filled triangles, dark $+$ fusicoccin; open triangles, light $\times$ fusicoccin. Data from Travis & Mansfield (1979a).

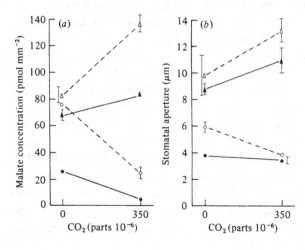

comparable effects observed in many other types of plant cell (Marrè, 1977).

The unexpected effect found in our experiments was a reversal of the carbon dioxide responses of stomata. Opening on isolated epidermis of *C. communis* treated with fusicoccin was markedly stimulated by carbon dioxide both in light and darkness, i.e. the reaction was the opposite of the normal one (Fig. 8). The same reversal of events was observed when total epidermal malate content was determined. In the absence of fusicoccin, the malate content was reduced as carbon dioxide concentration was increased, but with fusicoccin the opposite occurred, especially in light. The most intriguing aspect of these findings is that, in the presence of fusicoccin, malate contents and stomatal aperture respond in the manner we would expect if carboxylation of external carbon dioxide were making a contribution to malate formation and if opening were positively correlated with the amount of malate produced in the guard cells. We suggest the following interpretation.

(1) The familiar control of stomata by carbon dioxide is quite independent of any contribution that PEP carboxylation may make to the bulk of the guard cells' malate, namely that which balances potassium ion accumulation and is located in the vacuoles. There is another, unknown, effect of carbon dioxide on the processes involved in ion transport. This may result from acidification of the cytoplasm by malate produced from the carboxylation of carbon dioxide entering the guard cells from outside (see Raschke's proposals mentioned above) or it may be the result of a totally different action of carbon dioxide on transport processes.

(2) Fusicoccin enhances excretion of protons from the guard cells to such a degree that other controls over ion transport are overridden, one of which is that of carbon dioxide.

(3) The familiar stomatal response to carbon dioxide is thus absent, and one which is normally obscured is then revealed.

The fact that the carbon dioxide supply for malate formation appears to limit stomatal opening in the presence of fusicoccin is an indication that we should not rule out the possibility of some control over stomatal aperture at the level of anion formation in the guard cells. Our own results in Fig. 5 support the conclusion from other studies that the contribution by malate is variable, and can be moderated according to the availability of chloride ion. However, superimposed on this varying contribution was a remarkably consistent effect of both light and carbon dioxide on malate level (Fig. 6). This indicates that we must not neglect direct effects of these factors on anion formation; we may have to recognize another controlling influence at this level.

R. G. Jarvis acknowledges financial support from the Science Research Council and Shell Research Ltd.

References

Allaway, W. G. & Mansfield, T. A. (1967). Stomatal responses to changes in carbon dioxide concentration in leaves treated with 3-(4-chlorophenyl)-1, 1-dimethylurea. *New Phytologist* **66**, 57–63.

Allaway, W. G. & Milthorpe, F. L. (1976). Structure and functioning of stomata. In *Water Deficits and Plant Growth,* ed. T. T. Kozlowski, vol. IV, pp. 57–102. New York and London: Academic Press.

Balegh, S. E. & Biddulph, O. (1970). The photosynthetic action spectrum of the bean plant. *Plant Physiology* **46**, 1–5.

Davies, D. D. (1979). The central role of phosphoenolpyruvate in plant metabolism. *Annual Review of Plant Physiology* **30**, 131–58.

Graniti, A. & Turner, N. C. (1970). Effect of fusicoccin on stomatal transpiration in plants. *Phytopathologia Mediterranea,* **9**, 160–7.

Heath, O. V. S. & Russell, J. (1954). Studies in stomatal behaviour, VI: An investigation of the light responses of wheat stomata with the attempted elimination of control by the mesophyll. *Journal of Experimental Botany* **5**, 269–92.

Humble, G. D. & Hsiao, T. C. (1970). Light-dependent influx and efflux of potassium of guard cells during stomatal opening and closing. *Plant Physiology* **46**, 483–7.

Jarvis, R. G. & Mansfield, T. A. (1980). Reduced stomatal responses to light, carbon dioxide and abscisic acid in the presence of sodium ions. *Plant, Cell and Environment* **3**, 279–83.

Kuiper, P. J. C. (1964). Dependence upon wavelength of stomatal movement in epidermal tissue of *Senecio odoris. Plant Physiology* **39**, 952–5.

MacRobbie, E. A. C. (1977). Functions of ion transport in plant cells and tissues. In *International Review of Biochemistry, Vol. 13 Plant Biochemistry,* part ed. D. H. Northcote, pp. 211–47. Baltimore: University Park Press.

Mansfield, T. A. (1976). Delay in the response of stomata to abscisic acid in CO_2-free air. *Journal of Experimental Botany* **27**, 559–64.

Mansfield, T. A. & Meidner, H. (1966). Stomatal opening in light of different wavelengths: Effects of blue light independent of carbon dioxide concentration. *Journal of Experimental Botany* **17**, 510–21.

Marrè, E. (1977). Effects of fusicoccin and hormones on plant cell membrane activities: Observations and hypotheses. In *Regulation of Cell Membrane Activities in Plants,* ed. E. Marrè & O. Ciferri, pp. 185–202. Amsterdam: Elsevier/North-Holland Biomedical Press.

Neales, T. F. (1970). Effect of ambient carbon dioxide concentration on the rate of transpiration of *Agave americana* in the dark. *Nature* **228**, 880–2.

Nelson, S. D. & Mayo, J. M. (1975). The occurrence of functional nonchlorophyllous guard cells in *Paphiopedilum* spp. *Canadian Journal of Botany* **53**, 1–7.

Penny, M. G. & Bowling, D. J. F. (1974). A study of potassium gradients in the epidermis of intact leaves of *Commelina communis* L. in relation to stomatal opening. *Planta* **119**, 17–25.

Raghavendra, A. S. & Das, V. S. R. (1972). Control of stomatal opening by cyclic photophosphorylation. *Current Science* **41**, 150–1.

Raschke, K. (1975a). Simultaneous requirement of carbon dioxide and abscisic acid for stomatal closing in *Xanthium strumarium* L. *Planta* **125**, 243–59.

Raschke, K. (1975b). Stomatal action. *Annual Review of Plant Physiology* **26**, 309–40.

Raschke, K. (1977). The stomatal turgor mechanism and its responses to CO_2 and abscisic acid: observations and a hypothesis. In *Regulation of Cell Membrane Activities in Plants*, ed. E. Marrè & O. Ciferri, pp. 173–83. Amsterdam: Elsevier/North-Holland Biomedical Press.

Raschke, K. (1979). Movements of stomata. In *Encyclopedia of Plant Physiology, New Series, vol. 7*, ed. W. Haupt & M. E. Feinleib, pp. 383–441. Berlin: Springer-Verlag.

Raschke, K. & Dittrich, P. (1977). ^{14}C Carbon-dioxide fixation by isolated leaf epidermes with stomata closed or open. *Planta* **134**, 69–75.

Raschke, K. & Fellows, M. P. (1971). Stomatal movement in *Zea mays*: shuttle of potassium and chloride between guard cells and subsidiary cells. *Planta* **101**, 296–316.

Raschke, K. & Schnabl, H. (1978). Availability of chloride affects the balance between potassium chloride and potassium malate in guard cells of *Vicia faba* L. *Plant Physiology* **62**, 84–7.

Richardson, C. H., Truter, M. R., Wingfield, J. N., Travis, A. J., Mansfield, T. A. & Jarvis, R. G. (1979). The effect of benzo-18-crown-6, a synthetic ionophore, on stomatal opening, and its interaction with abscisic acid. *Plant, Cell and Environment* **2**, 325–7.

Rutter, J. C. & Willmer, C. M. (1979). A light and electron microscopy study of the epidermis of *Paphiopedilum* spp. with emphasis on stomatal ultrastructure. *Plant, Cell and Environment* **2**, 211–19.

Smith, H. (1975). Phytochrome and Photomorphogenesis. Maidenhead: McGraw Hill Book Co. Ltd.

Squire, G. R. & Mansfield, T. A. (1972). Studies of the mechanism of action of fusicoccin, the fungal toxin that induces wilting, and its interaction with abscisic acid. *Planta* **105**, 71–8.

Travis, A. J. & Mansfield, T. A. (1977). Studies of malate formation in 'isolated' guard cells. *New Phytologist* **78**, 541–6.

Travis, A. J. & Mansfield, T. A. (1979a). Reversal of the CO_2-responses of stomata by fusicoccin. *New Phytologist* **83**, 607–14.

Travis, A. J. & Mansfield, T. A. (1979b). Stomatal responses to light and CO_2 are dependent on KCl concentration. *Plant, Cell and Environment* **2**, 319–323.

Van Kirk, C. A. & Raschke, K. (1978). Presence of chloride reduces malate production in epidermis during stomatal opening. *Plant Physiology* **61**, 361–4.

Weyers, J. D. B. & Hillman, J. R. (1979). Uptake and distribution of abscisic acid in *Commelina* leaf epidermis. *Planta* **144**, 167–72.

Willmer, C. M. & Mansfield, T. A. (1969). A critical examination of the use of detached epidermis in studies of stomatal physiology. *New Phytologist* **68**, 363–75.

Willmer, C. M. & Mansfield, T. A. (1970). Effects of some metabolic inhibitors and temperature on ion-stimulated stomatal opening in detached epidermis. *New Phytologist* **69**, 983–92.

Wilson, J. A., Ogunkanmi, A. B. & Mansfield, T. A. (1978). Effects of external potassium supply on stomatal closure induced by abscisic acid. *Plant, Cell and Environment* **1**, 199–201.

R. LÖSCH & J. D. TENHUNEN

Stomatal responses to humidity – phenomenon and mechanism

The occurrence of stomatal responses to humidity

Stomata act as variable resistances in the soil–plant–atmosphere continuum (Weatherley, 1976), thus having a controlling influence on plant water status. On the other hand, they are influenced simultaneously by plant water potential. Stomata of plants suffering from water stress normally respond by closing to some degree (Stålfelt, 1956; Ketellapper, 1963). This is partly the result of general turgor loss in the leaf which then affects the guard cells, but in addition, there is a hormonal control mechanism in which water stress induces abscisic acid formation (Wright & Hiron, 1969; Kriedemann & Loveys, 1974). Raschke (1976) has hypothesized that this abscisic acid is transported to the guard cells, alters plasmalemma transport of H^+, and through the coupling of H^+ and K^+ transport leads to reduced solute content of the guard cells and closure. These reactions of the plant to water loss have been described in terms of a feedback control loop (Raschke, 1979). The feedback mechanism begins functioning after a threshold water potential has been surpassed (Hsiao, 1973; Beadle, Turner & Jarvis, 1978), and prevents further decrease in leaf water content. The threshold value (discussed further below) varies both with plant material and with environmental conditions and is apparently tuned to prevent excessive stress in a given situation.

However, a stomatal response to air humidity also occurs which may increase the efficiency of plant use of water and which is independent of the bulk water status of the leaf. In this case, stomata act as 'humidity sensors' (Ziegler, 1967; Lange, 1969, 1972). This behaviour has been termed 'feed-forward regulation' (Cowan, 1977; Cowan & Farquhar, 1977; Farquhar, 1978). With the feedforward controlling system, an immediate change in conductance occurs in response to a change in ambient evaporative conditions. The conductance change is independent of a change in the main component of transpiration although the transpiration rate is necessarily thereby affected and in many cases is reduced as evaporative demand increases. In contrast, feedback regulation can never result in a reduction of transpiration as evaporative demand increases. At best a constant transpiration rate can be maintained.

The existence of a direct response of stomata to changes in ambient humidity has been inferred from the results of experiments with well-watered plants in which stomatal resistance was recorded while the average water content of the ambient air was varied in otherwise constant environmental conditions (Rich & Turner, 1972; Davies & Kozlowski, 1974). These whole-leaf data showed that the stomata closed as the water saturation deficit between leaf and air (D (mg H_2O dm^{-3} of air)) increased. This could be explained by an immediate and direct response of the guard cells to atmospheric humidity or, alternatively, by failure of the internal water supply of the plant to keep up with local demand at the mesophyll evaporation sites, because of resistances in the conducting tissues. The local decrease in water potential would then result in stomatal closure. The experiments of Schulze *et al.* (1972) lend support to the explanation in terms of direct humidity response. Simultaneous measurements of leaf stomatal resistance and water content clearly demonstrated that stomata may open or close even though leaf water content is maintained constant. These observations support the idea of a direct humidity response, as do microscopic observations of isolated epidermal strips exposed on the inner surface to saturated air (Lange, Lösch, Schulze & Kappen, 1971). Adjacent groups of stomata in the epidermal strips respond by closing or opening, respectively, when exposed alternately to dry and moist air streams.

Fig. 1. Time course of stomatal opening and closing for two groups of stomata lying side-by-side within the same epidermal strip of *Polypodium vulgare* L. under the influence of alternating moist and dry air streams on the outside of the epidermis. Water vapour concentration at the inner side of the epidermis was at saturation; the vapour saturation deficit of the air passing over the epidermis was produced with silica-gel-dried (downward arrow) and vapour-saturated (upward arrow) air. Temperature of the epidermis, 26 °C; quantum flux density, 20 klx; CO_2-free air. Each point is the mean of four measurements. From Lange *et al* (1971).

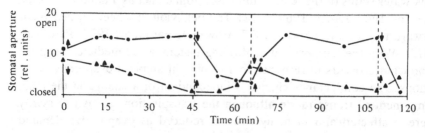

The time course of stomatal opening and closing in *P. vulgare* epidermal strips, in response to alternating treatment with dry and moist air streams, shown in Fig. 1, illustrates the humidity response phenomenon. Closure is about twice as fast as opening. Speed of closure is also influenced by the magnitude of the imposed change in ambient humidity (see also Lösch, 1977). Further variation in velocity of stomatal response may result from acclimation to prevailing humidity conditions during growth. For example, *Citrus* stomata become less sensitive to humidity after several days' exposure to step-wise decreasing humidities (Hall, Camacho-B & Kaufmann, 1975).

The results shown in Fig. 2 for closely related species within the genus *Sempervivum* s.l., endemic to the Macaronesian Islands, illustrate the breadth of range of stomatal response to change in ambient humidity. Rate of stomatal closure in isolated epidermal strips in response to dry air ranged from 0.04 to 1.08 μm min^{-1} Such differences are undoubtedly dependent on the structural as well as biochemical peculiarities of each epidermis.

Other observations suggest that the humidity response is even more intriguing and more complicated. *Aeonium glutinosum* from Madeira opens its stomata when a stream of dry air is blown across the outer surface of the isolated epidermis (Fig. 2). The reason for this behaviour is unknown.

Fig. 2. Time course of stomatal aperture for epidermal strips of several species of the genus *Sempervivum* s.l. when exposed to silica gel-dried-air at time zero. Air before time zero was vapour saturated at a temperature of 26 °C; quantum flux density, 20 klx; vapour saturation at the epidermal inner walls, CO_2-concentration that of normal air. Final measurement period showing stomatal re-opening for *Aeonium spathulatum* and *A. urbicum* and for *Aichryson laxum*, *A. villosum* and *A. dumosum* is in response to re-introduction of water vapour-saturated air.

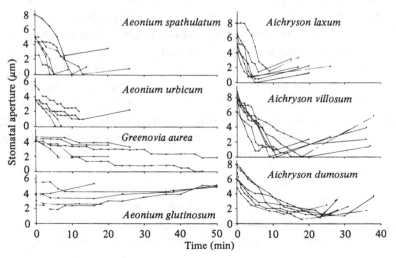

Similarly, Osmond *et al.* (1979) found stomatal closure in *Opuntia inermis*, a plant showing crassulacean acid metabolism (CAM), when it was exposed to dry air during the night, but no response or even a decrease in cladode resistance in the light. Two other studies of stomatal sensitivity to humidity in CAM plants (Conde & Kramer, 1975; Lange & Medina, 1979) have shown an increase in resistance or a decrease in carbon dioxide gas exchange in dry air in the light and in the dark. This appears to be less pronounced in the light (Lange & Medina, 1979). While it is well-established that the tuning of stomatal response to humidity is readily modified, the manner in which such change occurs is at present not understood and future work on the biochemical mechanisms involved is required.

At least 70 species of vascular plants have been shown to have a feedforward response of stomata to humidity. Lists of such species have been compiled by Sheriff (1977*b*) and by Lösch (1979*b*). They include ferns, gymnosperms, dicotyledons, and monocotyledons. The feedforward response has most often been found in C_3 plants, but it has also been reported for C_4 and CAM plants. The stomata of some plants seem to be non-reactive to changes in ambient humidity (e.g. *Zea mays* L.: Raschke & Kühl, 1969; *Desmodium uncinatum* DC.: Sheriff & Kaye, 1977; and others). We know of no data for mosses or the liverwort *Anthoceros*.

The influence of other factors on the humidity response of stomata

In the following sections, we discuss the influence of other environmental variables on the stomatal response to humidity. We have used the results obtained under controlled conditions with epidermal strips of *P. vulgare* to illustrate the effects generally observed. The results of these laboratory experiments seem to be generally true for other species; they demonstrate that steady-state stomatal aperture is not simply the result of the summed effects of single factors and allow us to construct a model which can be used to describe changes in stomatal aperture in response to change in any single variable and which also correctly describes the response to simultaneous change in several environmental factors under natural conditions.

Light and humidity

As a rule, stomata of C_3 and C_4 plants open in the light and close in the dark but CAM plants show the opposite behaviour. Among C_3 and C_4 plants there are species-specific differences in stomatal opening in response to light (Meidner & Mansfield, 1968) but it is generally accepted that stomatal opening saturates at rather low quantum flux density so that light saturation is reached in the early morning hours when humidity is still high (Meidner & Mansfield, 1968; Davies & Kozlowski, 1974).

Under natural conditions response to humidity is most often important in saturating light. Laboratory measurements at low quantum flux densities may give a false impression of field performance, since Kaufmann (1976) observed in *Picea engelmannii* (Parry) Engelm. that high light intensities reduce stomatal sensitivity to ambient humidity. Furthermore, *Fraxinus americana* L. and *Acer saccharum* Marsh, attain a new steady-state of stomatal resistance more rapidly when an abrupt change in humidity occurs at high rather than at low quantum flux densities (Davies & Kozlowski, 1974). Sheriff (1979) suggested that these altered sensitivities result from increased temperatures at high quantum flux densities, so that the strong light effect is indirect and mediated by improved hydration of the epidermis as a result of distillation from the warmer mesophyll tissue to the cells cooled by cuticular transpiration. Further investigation under isothermal conditions is required to determine whether there is an additional direct light effect on sensitivity to humidity.

Longer term effects of light in altering leaf structure are probably of considerable importance in modifying stomatal response to humidity. Davies & Kozlowski (1974) have speculated that much of the difference in stomatal sensitivity to humidity found in woody angiosperms growing on sites with differences in light climate may result from differing leaf structures. Differences in cuticular thickness and/or structure occur in sun and shade leaves of the same species. In the shrub *Hyptis emoryi* Torr. (Labiatae), sun leaves have higher resistances than shade leaves under the same humidity conditions (Smith & Nobel, 1977). It is possible that such differences balance water use at different canopy levels and co-ordinate activities in different microenvironments. Structural differences may also contribute to different humidity sensitivities in ab- and adaxial stomata. Aston (1976) showed that the stomata on the upper surface of sunflower leaves were much less responsive than the abaxial stomata. On the other hand, these results might be taken to indicate a direct effect of quantum flux density on the sensitivity of the humidity response. When sunflower leaves were turned over (Aston, 1978), the re-oriented abaxial stomata responded more like adaxial stomata. Changes in stomatal density of leaves developed under different light regimes may also contribute to changes in the humidity response, but these have not been studied.

Temperature and humidity

Fig. 3 shows the effects of temperature and humidity on stomata in epidermal strips of *P. vulgare*. At low saturation deficits, the stomata are open and they open further as temperature increases until at very high temperatures there is a slight stomatal closure (Lösch, 1977). At high temperatures, stomata are less sensitive to changes in humidity than at low temperatures (Hall & Kaufmann, 1975a).

The interaction between temperature and humidity is determined by plant growth history. The response surface shown in Fig. 3 can be shifted along the temperature axis if plants are grown under other temperature regimes (Lösch, 1977; Schulze *et al.*, 1974).

Carbon dioxide concentration, temperature, and humidity

Fig. 4 shows the influence of carbon dioxide concentration and vapour saturation deficit on stomatal aperture of *P. vulgare* at three temperatures. The effect of carbon dioxide concentration on pore size was larger

Fig. 3. Stomatal aperture of *Polypodium vulgare* as a function of temperature and humidity. Data from treatments of epidermal strips with vapour saturation in the substomatal air space in CO_2-free air, quantum flux density, 20 klx; plants grown in the greenhouse at 15 °C ±2 °C and 60% ±10% relative humidity. From Lösch (1977).

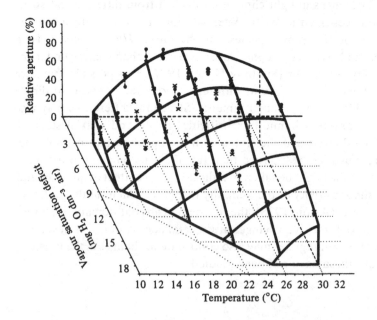

as temperature increased above 18 °C, while below 18 °C there was essentially no effect. However, stomatal aperture was reduced only about 30% at the most as carbon dioxide concentration was increased from zero to 400 cm³ m⁻³. The effect of carbon dioxide was essentially the same at all humidities. In contrast, Hall & Kaufmann (1975*b*) found in whole plants of *Sesamum indicum* L. that carbon dioxide concentration has a greater effect on stomatal resistance in dry air than in humid air.

Bulk water potential and humidity

If the air space below the epidermal strip is not saturated with water vapour, stomatal apertures are smaller than in conditions of water vapour saturation at the inner walls. We have interpreted this closure in response to an unsaturated air space as similar to that which occurs as a result of a drop in water potential, i.e. a feedback affecting general turgor. Fig. 5 shows that the effect of water potential on closure is strongly dependent on temperature and humidity. Reduction in pore size is almost negligible if water potentials in the sub-stomatal air space do not drop below −0.8 MPa and if

Fig. 4. Stomatal aperture of *Polypodium vulgare* as a function of CO_2-concentration, humidity and temperature. Other conditions as described for Fig. 3, except cultivation conditions of the plants: 20 °C ± 2 °C and 80% ± 10% relative humidity.

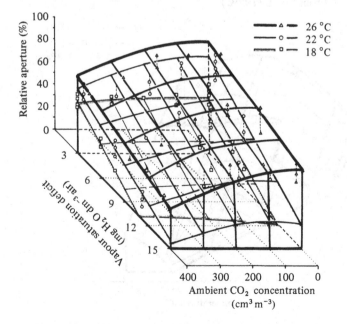

temperatures are low. At higher temperatures there is a tendency for all decreases in water potential to promote stomatal closure. In other words, there is a water potential threshold at low temperature which decreases and finally disappears as temperature increases. Fig. 5 also shows that both feedback stomatal closure in response to plant water deficit and feedforward response to prevailing air humidity function concurrently as proposed by Raschke (1979).

Information on the combined action of bulk water potential and ambient humidity on stomata in whole plants is limited, and studies which include temperature are scarce indeed. Fetcher (1976) and also Running (1976) found stomatal apertures in conifers to be governed by ambient humidity when water status was favourable whereas, at very low water potentials, stomata remained closed even in moist air. In between, ambient humidity and plant water potential both appeared to have an effect. The increased effect of water potential below a threshold has been observed in many species (Hsiao, 1973) but not in others, e.g. *Corylus avellana* L. (Schulze & Küppers, 1979). The results with *P. vulgare* suggest that observation of a threshold value may depend on the temperature dependency of the threshold for the particular species and the temperatures obtaining during experimentation. As with *P. vulgare*, Lawlor & Lake (1976) found for

Fig. 5. Stomatal aperture of *Polypodium vulgare* as a function of temperature, humidity, and several vapour saturation deficits at the epidermal inner walls (expressed as water potential); other conditions see Fig. 3. From Lösch (1979*a*).

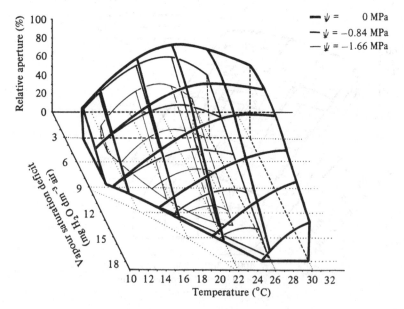

Lysimachia nummularia L., *Lolium perenne* L., and *Trifolium repens* L. that sensitivity of stomata to water potential was less in dry air than in moist air. Schulze & Küppers (1979, Figure 5B) showed that the sensitivity of stomata of *C. avellana* to change in humidity decreased as water potential decreased.

A parallelism seems to exist in the stomatal sensitivity to leaf bulk-water potential and to air humidity. Sheriff & Kaye (1977) found a high stomatal responsiveness to change in humidity in *Macroptilium atropurpureum* (DC.) Urb. in controlled environment studies. In contrast, stomata of *Desmodium uncinatum* were unresponsive. Under field conditions, Ludlow & Ibaraki (1979) demonstrated that stomata of *M. atropurpureum* closed rapidly in response to a drop in leaf water potential between -0.5 and -1.0 MPa, while the stomata of *D. uncinatum* normally remained open even at low water potentials. Daily time-course data for *M. atropurpureum* showed time periods during which humidity appeared to be the main factor affecting stomatal resistance and periods in which water potential appeared to play a dominant role but simultaneous changes in quantum flux density and leaf temperature made it difficult to sort out the interplay of water potential and humidity.

Modelling of stomatal behaviour under the combined influence of air humidity and other relevant factors

Jarvis (1976) proposed a general model for stomatal behaviour under natural conditions. He emphasized that to apply the model to a particular species, measurements should be made over a wide range of values of the relevant variables to determine parameter values without bias. As illustrated in Fig. 6 this is difficult to achieve in the field where quantum flux density, temperature, and vapour saturation deficit are very closely coupled. Nevertheless, Jarvis determined model parameters for *Picea sitchensis* (Bong.) Carr. and *Pseudotsuga menziesii* (Mirb.) Franco using data from measurements of leaf gas exchange. He attributed substantial differences in the numerical values of the parameters to species differences and to phenological differences between the experimental trees.

We have used the data already illustrated in Figs. 3 and 5 in conjunction with other measurements of stomatal response, to express mathematically steady-state stomatal aperture as a function of humidity, temperature, leaf water potential, and quantum flux density. An advantage of basing a model on these data from *P. vulgare* is that interactions between variables which were not included in Jarvis's model can be included. For simplicity most relationships in the model are described with linear functions.

The relationship between aperture and temperature at $D = 0$ and water potential = 0 (Fig. 3), can be described empirically by two linear equations. Temperature induces stomatal opening until an optimum is reached (eqn 1A in Table 1) and then induces closure (eqn 1B). Increasing D reduces aperture at any temperature and is described by another linear relation. An interaction (eqn 2) exists between the effects of temperature and D since the slope of the regression between aperture and D decreases as temperature increases. Eqn 3 in Table 1 combines these ideas and describes the surface of Fig. 3.

Eqn 4 describes the temperature dependency of the threshold water potential (ψ_{min}) shown in Fig. 5 and eqn 5 calculates the effective range of water potential beyond this threshold over which closure occurs (ψ_{eff}). In $P.$ $vulgare$ the sensitivity of stomatal closure to water potential decreases as D increases and this interaction is accounted for by eqn 6. Eqn 7 combines these water potential effects with the effects of temperature and D.

We assume in the model that light determines the maximal stomatal aperture attainable under otherwise optimal conditions for opening and that this light effect implicitly includes effects of carbon dioxide. Some variation in carbon dioxide concentration between 200 and 300 cm^3 m^{-3} at a constant quantum flux density had little effect on stomata of $P.$ $vulgare$ (Fig. 4) or $Prunus$ $armeniaca$ L. to which the model is also to be applied (O. L. Lange & A. Meyer, unpublished data). Further separation of the effects of light and carbon dioxide concentration have not been considered. Eqn 8 describes the

Fig. 6. Distributions of combinations of two environmental variables. (a). Random scatter of points that allows bias-free estimation of both temperature and humidity effects (Jarvis, 1976). (b). Combinations applied in laboratory experiments with epidermal strips of $Polypodium$ $vulgare$ to study stomatal behaviour with respect to both factors; the point-free corner (bottom left) is the result of variable combinations which lead to totally closed stomata. (c). Combinations within the value range of case b occurring naturally during six days of the vegetation period in 1976 in which the general weather conditions varied greatly. Gas-exchange measurements with $Prunus$ $armeniaca$. From Lange & Meyer (1979).

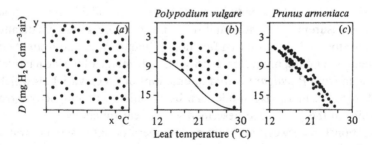

effect of light on stomatal aperture with a hyperbolic function (Turner & Begg, 1973; Tenhunen & Gates, 1975). The parameter x_{max} is the maximum attainable aperture of a stoma and allows for species-specific differences in stomatal dimensions. Limitations to the aperture attributable to temperature, D, and water potential (eqn 7) are then determined relative to the maximum aperture established for a prevailing quantum flux density (eqn 9).

Table 1. *Model designed to describe stomatal relative aperture, conductance, and resistance as a function of temperature (T_1), vapour pressure deficit (D), bulk water potential (ψ), and quantum flux density (Q), based on data from epidermal strips of* Polypodium vulgare.

Relative aperture $= f(T_1)$		
$x_1 = a_1 + b_1 T_1$	$T_1 <$ optimum temperature	(1A)
$x_1 = a_2 + b_2 T_1$	$T_1 >$ optimum temperature	(1B)
Sensitivity to D $= f(T_1)$ (Interaction term)		
$S_1 = c - d\,T_1$		(2)
Relative aperture $= f(T_1,D)$		
$x_2 = x_1 - S_1 D$	$\psi = 0$	(3)
Threshold water potential (Interaction term)		
$\psi_{min} = e T_1 - f$	below $T_1 = f/e$	(4A)
$\psi_{min} = 0$	above $T_1 = f/e$	(4B)
Effective water potential		
$\psi_{eff} = \left\| \psi - \psi_{min} \right\|$	$\left\|\psi\right\| > \left\|\psi_{min}\right\|$	(5A)
$\psi_{eff} = 0$	$\left\|\psi\right\| < \left\|\psi_{min}\right\|$	(5B)
Sensitivity to water potential $= f(D)$ (Interaction term)		
$S_2 = g - hD$		(6)
Relative aperture $= f(T_1, D, \psi)$		
$x_3 = x_2 - S_2 \psi_{eff}$		(7)
Maximal aperture $= f(Q)$		
$x_4 = x_{max}/(1 + k_1/Q)$		(8)
Actual aperture $= f(T_1, D, \psi, Q)$		
$x_5 = x_4 x_3/100$		(9)
Conductance $= f$ *(actual aperture of a single stoma)*		
$G = G_{max}/(1 + k_2/x_5)$		(10)
Resistance $= f$ *(conductance)*		
$R = 1/G$		(11)

Units of variables
T_1 is the leaf temperature (°C).
D is the water vapour saturation deficit between leaf and air (mg H_2O dm^{-3} air).
ψ is the xylem water potential of the leaf (MPa).
Q is the quantum flux density (kilolux).
x_1, x_2, x_3 is the relative aperture (%).
x_4, x_5, x_{max} is stomatal aperture (μm^2).
G, G_{max} is leaf conductance (cm s^{-1}).
R is leaf resistance (s cm^{-1}).

The relationship between leaf conductance and aperture of a single stoma can be described by another hyperbolic relationship (eqn 10) (Cowan & Milthorpe, 1968) in which the parameter k_2 is the area of opening resulting in a conductance equal to one-half of the maximum conductance G_{max}. For lack of better information, we have assumed that the relation between pore area and conductance for wheat given by Cowan & Milthorpe can be applied to *P. vulgare* and *Prunus armeniaca*. In practice, the relation is only slightly non-linear. Finally, leaf diffusion resistance is obtained as the reciprocal of conductance (eqn 11).

Fig. 7 illustrates the three-dimensional surface obtained for leaf resistance of *P. vulgare*. Comparison between Figs. 5 and 7 shows that there are regions in which large changes in aperture have almost no effect on leaf resistance because of the extreme non-linearity of eqn 11.

Since the stomatal behaviour of *P. vulgare* is consistent with that of a variety of species, we have used the model to describe leaf resistance in other species. For any particular application, values of the species-specific parameters must be obtained, because overall sensitivity of the stomata depends on environmental conditions during growth and species-specific anatomical and biochemical characteristics, e.g. Davies (1977).

Extensive data for leaf gas exchange in *Prunus armeniaca* were kindly provided by A. Meyer to allow a test of the model (see Lange & Meyer, 1979). The data consisted of daily time-course measurements of stomatal resistance, T_1, ψ and Q over 40 days throughout the entire vegetation period in 1976. We assumed in the analysis that the continuously changing combi-

Fig. 7. Hypothetical leaf resistance of *Polypodium vulgare* as a function of temperature, humidity, and water potential as simulated by the model.

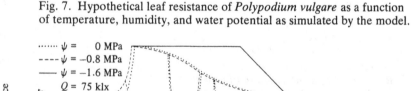
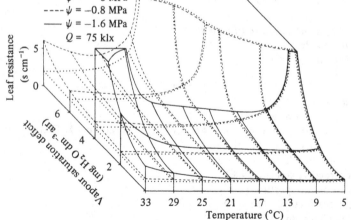

nations of environmental variables produced a sequence of near steady-state leaf resistances. Using data collected on three days with extremely different weather conditions, model parameters for apricot were obtained: in general, sensitivity to D and to water potential was less than originally found in *P. vulgare*. Using appropriate parameters and conditions, about three-quarters of the daily time courses were simulated satisfactorily. Fig. 8 shows surfaces describing stomatal resistance of apricot as a function of temperature and D, at a saturating quantum flux density of 75 klx, and for leaf water potentials of 0, −2.0 and −4.0 MPa. From these surfaces it is apparent how important the interplay of humidity and leaf temperature is in determining resistance in *Prunus armeniaca*. Water potential plays a lesser role but is an important modifier of the temperature and humidity responses. This agrees with earlier field observations (Schulze *et al.*, 1975*b*) which indicated that stomatal resistance in apricot is only loosely coupled to leaf bulk water potential.

The mechanism underlying stomatal response to humidity

Lange *et al.* (1971) proposed that peristomatal transpiration provides a physiological basis for the response of stomata to humidity. This concept, first developed by Seybold (1961/62), is based on the idea that

Fig. 8. Predicted leaf resistance of *Prunus armeniaca* as a function of temperature, humidity, and water potential, at high quantum flux density (75 klx) simulated by the model after determination of the parameters from the gas exchange data of Lange & Meyer (1979).

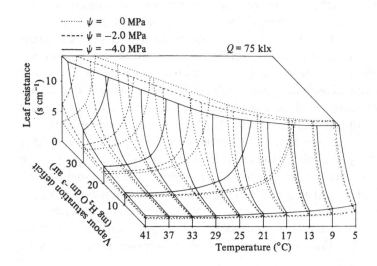

there is a cuticular transpiration flux from the cell walls of the stomatal apparatus which is larger than that from the cell walls of other epidermal cells.

Histochemical data lend support to the existence of peristomatal transpiration (Maier-Maercker, 1979a). Chemical tracers appear to accumulate at endpoints of the transpiration stream, i.e. where evaporation occurs (Crowdy & Tanton, 1970: lead-EDTA; Maercker, 1965a: TlCl; Aston & Jones, 1976: monosilicic acid; Burbano et al., 1976: Prussian blue). Deposits of these substances in the epidermis have been detected primarily at three locations: 1) along guard cell and subsidiary cell inner walls, 2) in the guard cell spicules, and 3) in regions where the outer surface of the epidermis intersects with the anticlinical walls of the guard cells. While these deposits might be the result of coincidental concentration of the tracers at these locations, direct and convincing evidence for more intense transpiration at these points is seen in the autoradiographic experiments with tritiated water by Maercker (1965a,b) and by hydrophotography (Maier-Maercker, 1979a). Special cytological means of channelling water such as ectodesmata (Franke, 1967) are not required: less cutinized regions alone could be responsible (Sheriff, 1977a). Edwards & Meidner (1978) obtained electron micrographs of sections of P. vulgare guard cells which show uncutinized areas, especially along the guard cell inner walls where histochemical investigations indicated the greatest water loss. Sheriff (1977a) found that peristomatal transpiration from both the inner epidermal walls and the outer walls was significant in total water loss from epidermal strips of Tradescantia: stomatal closure in response to dry air was reduced by one-half when evaporation from the inner walls was prevented.

In the simplest scheme of peristomatal transpiration the enhanced loss of water from the guard cells when humidity decreases leads to a decrease in turgor pressure of the guard cells and results in stomatal closure. Closure is caused by a change in pressure exerted on the guard cell walls, since the turgor of adjacent epidermal cells decreases to a lesser extent. The guard cells apparently sense changes in ambient humidity because the vapour pressure gradients are altered at the sites of evaporation from the guard cells. The guard-cell vacuoles are closely linked to evaporation from the cell walls. Water molecules moving along the epidermal pathway will either move into the guard cells or to evaporation sites in the walls depending on the relative size of the water potential gradients. The very low osmotic potentials which exist in guard cell vacuoles are vital to maintain turgor competitively (Meidner & Edwards, 1975). Direct turgor measurements indicate that the turgor changes needed to effect guard cell movement are relatively small, so that a low osmotic potential is necessary to counteract the tendency for water to flow toward the low matric potentials in the guard cell walls. When the ambient humidity is decreased the wall matric potentials are lowered further and water is withdrawn from the guard cell vacuoles leading to reduction in guard cell turgor and narrowing of the stomatal pore.

Maier-Maercker (1979*a*) has analysed stomatal closure which results from a sudden decrease in ambient humidity in the context of a general supply–demand scheme considering water potentials in three compartments: guard cells, subsidiary cells, and sites of evaporation. She concluded that the reduction in aperture brought about initially by increase in water loss as a result of increase in the rate of evaporation, can only be maintained

Fig. 9. Time course of stomatal aperture in isolated epidermal strips of *Valerianella locusta* L. and corresponding guard cell potassium content (as determined by Macallum's stain) during a change from high (D, 2 mg H_2O dm^{-3} air) to low ambient humidities (D, 7 mg H_2O dm^{-3} air). The change in guard cell potassium content lags behind change in stomatal aperture. Data from Lösch & Schenk (1978). Temperature = 26 °C; water vapour-saturated air below the epidermis.

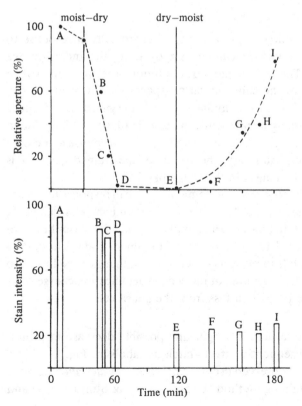

in the steady state through a decrease in the guard cell solute content. Such changes could be achieved in several ways.

Potassium content. Studies in our laboratory in which the potassium content of *Valerianella locusta* guard cells was measured, have confirmed Maier-Maercker's contention that 'the solute contents of the guard cells were reduced in a "follow-up" system' (Maier-Maercker, 1979*b*, p. 168). Fig. 9 shows that when stomata close in response to dry air, the guard cell potassium content remains initially unchanged, but after most of the change in aperture has taken place, the potassium content decreases (see also Lösch & Schenk, 1978; Lösch, 1978).

Malate. Some preliminary experiments suggest a change in guard cell malate content during humidity induced movement but further methodological difficulties must be overcome before clear conclusions can be drawn about the metabolism of this possible counter-ion for potassium. We have, however, observed changes in guard cell pH which may be linked to malate metabolism.

pH. We have followed pH changes of guard cells in response to changes in humidity in *Valerianella locusta* by using the indicator dye bromocresol purple. The dye in aqueous solution was injected with a hypodermic syringe into the substomatal air spaces of leaves on the day before an experiment. The dye accumulates solely in the guard cells and does not impair their functioning if sufficiently diluted (1:4000). We have found that as the stomata of intact leaves opened and closed in response to increase and decrease in ambient humidity, the colour of the stained guard cells changed, indicating pH changes in the range pH 5–7.

Fig. 10 shows that after the onset of stomatal closure in response to dry air, guard cell pH may or may not at first decrease; it then increases to a peak, decreases a second time, increases to a second peak, and decreases finally to a value less than or equal to that observed in open stomata. These data indicate considerable shift in proton concentrations, that may be linked to potassium transport. The beginning of the second acidification course coincides with the onset of potassium loss from the guard cells.

Starch. The initial loss in turgor and possible decrease in proton concentration may subsequently affect starch metabolism. Fig. 11 shows changes in starch content observed in epidermal strips of *Valerianella locusta* following the iodine staining method (Urbach, Rupp & Sturm, 1976) and

Fig. 10. Time course of stomatal opening and closing in response to change in humidity from vapour-saturated to silica gel-dried air in isolated epidermal strips of *Valerianella locusta* and corresponding alternating colour changes in the indicator dye, bromocresol purple, accumulated in the guard cells (low relative units, pH low; high relative units, pH high). Temperature = 26 °C, water vapour-saturated air below the epidermis.

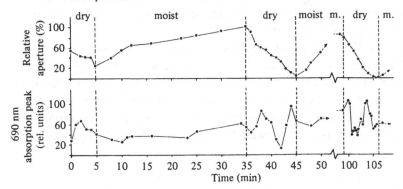

Fig. 11. Time course of stomatal opening and closing in response to change in humidity from vapour-saturated to silica gel-dried air in isolated epidermal strips of *Valerianella locusta* and corresponding changes in guard cell starch contents. Each of the six treatments was terminated by iodine-staining of the epidermis at the end of the respective humidity treatment. Temperature = 26 °C, water vapour-saturated air below the epidermis.

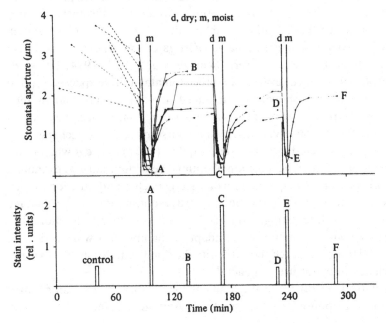

the scoring system of Yemm & Willis (1954). At times A to F, starch was built up and also very quickly broken down when the epidermal strips were treated with several cycles of moist and dry air. The speed with which starch was built up during these stomatal closing movements in response to changes in humidity is remarkable. In experiments in which well-watered whole plants were placed in a growth chamber and subjected to moist–dry–moist air treatments, a clear linear relationship between relative starch content and pore aperture was also observed.

Feedforward responses of stomata to humidity are associated with metabolic events. As in the case with other variables inducing stomatal movements, such as light, temperature, or carbon dioxide concentration, humidity-induced changes appear to influence the solute content of the guard cells. The humidity response of stomata thus appears to be a topic of study which will profit from more biochemical and biophysical cellular research.

Direct humidity influence on stomatal aperture, its ecological meaning and importance for the plant

Arid zone plants are subjected to selective pressure to conserve water and might be expected to optimize water use (i.e. obtain the greatest possible carbon gain per unit of water transpired). Cowan (1977) and Cowan & Farquhar (1977) have discussed the manner in which plants might sense the environment in order to optimize water use efficiency and they have demonstrated that a feedforward humidity response provides just such a means. Investigations of gas exchange of irrigated *Prunus armeniaca* and of native plants in the Negev Desert (Schulze *et al.*, 1972, 1974, 1975*a*, *b*; Lange *et al.*, 1975) have revealed that the stomata in these species respond strongly to humidity, showing pronounced mid-day closure and thereby decreasing the transpiration ratio (moles water transpired/moles carbon dioxide fixed). The observed daily time course of leaf gas exchange was similar to that predicted by Cowan & Farquhar (1977) to occur when water use is optimized. While the question of optimal behaviour must be examined more closely and may help guide further study of stomatal function in such plants, there is no doubt that the feedforward response confers an advantage on these species. Interestingly, the feedforward response to humidity has now been shown to function in an adaptive manner in a wide variety of habitats and future research should clarify the significance of such a response for efficiency of water use in general.

It is relatively seldom that the limits of species distribution can be attributed to a response to any single environmental factor. However, measurements of gas exchange and of microclimate of some tundra species (Johnson & Caldwell, 1976) suggested that a high stomatal sensitivity to humidity may be the main factor allowing survival on extreme sites. *Carex aquatilis* Wahlenb. and *Geum rossii* (R.Br.) Ser., which respond strongly to humidity, are widely distributed, while *Dupontia fischeri* R.Br. and *Deschampsia caespitosa* (L.) Beauv., which respond poorly to humidity, are restricted to wet sites. Ludlow & Ibaraki (1979) attributed successful growth of the tropical pasture legume *M. atropurpureum* in semi-arid areas of Australia to the very sensitive response of its stomata to both ambient humidity and plant water stress. Less sensitive co-occurring species are out-competed in the central area of distribution of the species.

Wind may be expected to increase the response of stomata to humidity by reducing the thickness of the boundary layer above the sites of peristomatal transpiration (Grace, 1977). In a particularly interesting study, stomatal sensitivity to windspeed in two morphological ecotypes of *Cytisus scoparius* L. (broom) was shown to be correlated with the microclimates of the habitats in which the ecotypes are successful (Davies, Gill & Halliday, 1978). The prostrate subspecies *maritimus* has a scattered distribution on exposed cliffs in the south-west of the British Isles. The stomata in the ecotype closed rapidly in response to abrupt changes in windspeed that altered evaporative conditions, whereas the stomata of the erect subspecies *scoparius*, which is confined naturally to more sheltered habitats, showed no such response.

Larcher (1975) speculated that the sensitivity of stomata to humidity has co-evolved with morphological adaptations that alter the humidity gradient in the boundary layer of leaves and plant canopies. For example, many plants growing at high altitudes in the Andean Paramos possess a dense indumentum, small leaves, cushion-like or dwarf-like habit, and other characteristics usually classified as xeromorphic. Since these plants appear to be well supplied with water even during the dry season, Larcher postulated that these adaptations create a favourable microclimate, an important aspect of which is to maintain high humidity around the leaves. Such adaptations would be effective in insulating humidity-sensitive stomata from the generally dry alpine air. Measurements of transpiration and stomatal conductance in *Loiseleuria procumbens* (L.) Desv., *Calluna vulgaris* (L.) Hull and other dwarf shrubs (Körner, 1975, 1977; Cernusca, 1976) growing above the timberline in the Alps agree with the view that plant growth form creates a microenvironment suitable for open stomata and that when the boundary layer is disturbed by wind, very rapid stomatal closure takes place.

The results obtained with different species of *Aichryson* (Fig. 2), support Larcher's hypothesis. The leaves of these plants are hirsute, but the density of hair covering is quite different in the three species (Table 2). When dry air was directed on to an individual stoma with a fine capillary, the most rapid stomatal closure occurred in the species with the most dense indumentum, i.e. in the leaf that would normally be most effectively insulated by a thick boundary layer. In this species, *A. laxum*, stomatal closure was four times slower when the dry air was blown over the leaf, so that the hairs were able to insulate the stomata from the dry air, than when the dry air stream was directed on to an individual stoma.

Conclusions

Direct responses of stomata to changes in humidity in the air are widespread. They are mediated by a very intricate mechanism that is extremely sensitive to change in the content and distribution of water in the leaf cells. Peristomatal transpiration is involved in the sensing of air humidity and guard cell metabolism in the response. Considerable further biochemical and physiological work is required before a satisfactory understanding of the response to humidity can be achieved. Since the response to humidity is very much modified by other variables to which the stomata respond, e.g. light, temperature and water potential, it will be interesting to see what interactions result when more than one variable affects the same metabolic events responsible for guard cell turgor. Failure to consider the effect of humidity on stomatal aperture has contributed in the past to difficulties in interpretation of field studies of leaf gas exchange. Investigations of the humidity response have allowed us to construct a model which, we hope, will better explain changes in stomatal resistance under natural conditions. In certain habitats where plants are exposed to extreme water stress, such as deserts, high mountains and windy coasts, the importance of a response to humidity has already been demonstrated. Studies of the importance of a response to humidity in other habitats, such as temperate forests, may

Table 2.

	Rate of stomatal closure (μm min^{-1})	Density of hairs on lower epidermis (mm^{-2})
Aichryson laxum	1.08	19
Aichryson villosum	0.32	6
Aichryson dumosum	0.16	4

provide us with very interesting examples of plant adaptation to a complex of changing light, temperature, and humidity patterns, modified by gradual seasonal change in water relations.

We thank Ms Laura Tenhunen for patient help in preparation of the text and Ms Wilma Samfass, Ms Claudia Büttner and Ms Doris Faltenbacher for technical help in preparation of the figures. We gratefully acknowledge the helpful suggestions of Professor Dr Otto L. Lange and of Professor P. G. Jarvis.

References

Aston, M. J. (1976). Variation of stomatal diffusive resistance with ambient humidity in sunflower (*Helianthus annuus*). *Australian Journal of Plant Physiology* **3**, 489–502.

Aston, M. J. (1978). Differences in the behaviour of adaxial and abaxial stomata of amphistomatous sunflower leaves: Inherent or environmental? *Australian Journal of Plant Physiology* **5**, 211–18.

Aston, M. J. & Jones, M. M. (1976). A study of the transpiration surfaces of *Avena sterilis* L. var. Algerian leaves using monosilicic acid as a tracer for water movement. *Planta* **130**, 121–9.

Beadle, C. L., Turner, N. C. & Jarvis, P. G. (1978). Critical water potential for stomatal closure in Sitka spruce. *Physiologia Plantarum* **43**, 160–5.

Burbano, J. L., Pizzolato, T. D., Morey, P. R. & Berlin, J. D. (1976). An application of the Prussian blue technique to a light microscope study of water movement in transpiring leaves of cotton (*Gossypium hirsutum* L.). *Journal of Experimental Botany* **27**, 134–44.

Cernusca, A. (1976). Energie- und Wasserhaushalt eines alpinen Zwergstrauchbestandes während einer Föhnperiode. *Archiv für Meteorologie, Geophysik und Bioklimatologie* Series B, **24**, 219–41.

Conde, L. F. & Kramer, P. J. (1975). The effect of vapour pressure deficit on diffusion resistance in *Opuntia compressa*. *Canadian Journal of Botany* **53**, 2923–6.

Cowan, I. R. (1977). Stomatal behaviour and environment. *Advances in Botanical Research* **4**, 117–228.

Cowan, I. R. & Farquhar, G. D. (1977). Stomatal function in relation to leaf metabolism and environment. In *Integration of Activity in the Higher Plant, Symposia of the Society for Experimental Biology*, **31**, 471–505. Cambridge University Press.

Cowan, I. R. & Milthorpe, F. L. (1968). Plant factors influencing the water status of plant tissues. In *Water Deficits and Plant Growth*, ed. T. T. Kozlowski, Vol. 1, pp. 137–193. New York and London: Academic Press.

Crowdy, S. H. & Tanton, T. W. (1970). Water pathways in higher plants. I. Free space in wheat leaves. *Journal of Experimental Botany* **21**, 102–12.

Davies, W. J. (1977). Stomatal responses to water stress and light in plants grown in controlled environments and in the field. *Crop Science* **17**, 735–40.

Davies, W. J., Gill, K. & Halliday, G. (1978). The influence of wind on the behaviour of stomata of photosynthetic stems of *Cytisus scoparius* (L.) Link. *Annals of Botany* 42, 1149–54.

Davies, W. J. & Kozlowski, T. T. (1974). Stomatal responses of five woody angiosperms to light intensity and humidity. *Canadian Journal of Botany* 52, 1525–34.

Edwards, M. & Meidner, H. (1978). Stomatal responses to humidity and the water potentials of epidermal and mesophyll tissue. *Journal of Experimental Botany* 29, 771–80.

Farquhar, G. D. (1978). Feedforward responses of stomata to humidity. *Australian Journal of Plant Physiology* 5, 787–800.

Fetcher, N. (1976). Patterns of leaf resistance to lodgepole pine transpiration in Wyoming. *Ecology* 57, 339–45.

Franke, W. (1967). Ektodesmen und die peristomatäre Transpiration. *Planta* 73, 138–54.

Grace, J. (1977). *Plant Response to Wind*. 204 pp. London, New York and San Francisco: Academic Press.

Hall, A. E., Camacho-B, S. E. & Kaufmann, M. R. (1975). Regulation of water loss by citrus leaves. *Physiologia Plantarum* 33, 62–5.

Hall, A. E. & Kaufmann, M. R. (1975a). Regulation of water transport in the soil-plant-atmosphere continuum. In *Perspectives of Biophysical Ecology*, ed. D. M. Gates & R. B. Schmerl. *Ecological Studies* 12, 187–202. Berlin, Heidelberg, New York: Springer-Verlag.

Hall, A. E. & Kaufmann, M. R. (1975b). Stomatal response to environment with *Sesamum indicum* L. *Plant Physiology* 55, 455–9.

Hsiao, T. C. (1973). Plant responses to water stress. *Annual Review of Plant Physiology* 24, 519–70.

Jarvis, P. G. (1976). The interpretation of the variations in leaf water potential and stomatal conductance found in canopies in the field. *Philosophical Transactions of the Royal Society of London* Series B, 273, 593–610.

Johnson, D. A. & Caldwell, M. M. (1976). Water potential components, stomatal function, and liquid phase water transport resistances of four arctic and alpine species in relation to moisture stress. *Physiologia Plantarum* 36, 271–8.

Kaufmann, M. R. (1976). Stomatal response of Englemann spruce to humidity, light and water stress. *Plant Physiology* 57, 898–901.

Ketellapper, H. J. (1963). Stomatal physiology. *Annual Review of Plant Physiology* 14, 249–70.

Körner, C. (1975). Wasserhaushalt und Spaltenverhalten alpiner Zwergsträucher. *Verhandlungen der Gesellschaft für Ökologie, Wien* pp. 23–30.

Körner, C. (1977). Blattdiffusionswiderstände verschiedener Pflanzen im alpinen Grasheidegürtel der Hohen Tauern. *Veröffentlichungen des Österreichischen MaB-Hochgebirgsprogrammes Hohe Tauern* 1, 69–81. Innsbruck: Universitätsverlag Wagner.

Kriedemann, P. E. & Loveys, B. R. (1974). Hormonal mediation of plant responses to environmental stress. In *Mechanisms of Regulation of Plant Growth*. ed. R. L. Bieleski, A. R. Ferguson & M. M. Creswell. *Bulletin of the Royal Society of New Zealand* 12, 461–5.

Lange, O. L. (1969). Wasserumsatz und Stoffbewegungen. *Fortschritte der Botanik* 31, 76–86.

Lange, O. L. (1972). Wasserumsatz und Stoffbewegungen. *Fortschritte der Botanik* **34**, 91–112.

Lange, O. L., Lösch, R., Schulze, E. D. & Kappen, L. (1971). Responses of stomata to changes in humidity. *Planta* **100**, 76–86.

Lange, O. L. & Medina, E. (1979). Stomata of the CAM plant *Tillandsia recurvata* respond directly to humidity. *Oecologia* **40**, 357–63.

Lange, O. L. & Meyer, A. (1979). Mittäglicher Stomataschluß bei Aprikose (*Prunus armeniaca*) und Wein (*Vitis vinifera*) im Freiland trotz guter Bodenwasser-Versorgung. *Flora, Jena* **168**, 511–28.

Lange, O. L., Schulze, E. D., Kappen, L., Buschbom, U., Evenari, M. (1975). Photosynthesis of desert plants as influenced by internal and external factors. In *Perspectives of Biophysical Ecology*, ed. D. M. Gates & R. B. Schmerl. *Ecological Studies* **12**, 121–43.

Larcher, W. (1975). Pflanzenökologische Beobachtungen in der Páramostufe der venezolanischen Anden. *Anzeiger der mathematisch-naturwissenschaftlichen Klasse der Österreichischen Akademie der Wissenschaften* 194–213.

Lawlor, D. W. & Lake, J. V. (1976). Evaporation rate, leaf water potential and stomatal conductance in *Lolium*, *Trifolium* and *Lysimachia* in drying soil. *Journal of Applied Ecology* **13**, 639–46.

Lösch, R. (1977). Responses of stomata to environmental factors – Experiments with isolated epidermal strips of *Polypodium vulgare*. I. Temperature and humidity. *Oecologia* **29**, 85–97.

Lösch, R. (1978). Veränderungen im stomatären Kaliumgehalt bei Änderungen von Luftfeuchte und Umgebungstemperatur. *Berichte der Deutschen Botanischen Gesellschaft* **91**, 645–56.

Lösch, R. (1979a). Responses of stomata to environmental factors – Experiments with isolated epidermal strips of *Polypodium vulgare*. II. Leaf bulk water potential, air humidity, and temperature. *Oecologia* **39**, 229–38.

Lösch, R. (1979b). Stomatal responses to changes in air humidity. In *Structure, Function, and Ecology of Stomata*, ed. D. N. Sen, D. D. Chawan & R. P. Bansal, pp. 189–216. Dehra Dun: Bishen Singh Mahendra Pal Singh.

Lösch, R. & Schenk, B. (1978). Humidity responses of stomata and the potassium content of guard cells. *Journal of Experimental Botany* **29**, 781–7.

Ludlow, M. M. & Ibaraki, K. (1979). Stomatal control of water loss in siratro (*Macroptilium atropurpureum* (DC)Urb.), a tropical pasture legume. *Annals of Botany* **43**, 639–47.

Maercker, U. (1965a). Zur Kenntnis der Transpiration der Schließzellen. *Protoplasma* **60**, 61–78.

Maercker, U. (1965b). Mikroautoradiographischer Nachweis tritiumhaltigen Transpirationswassers. *Die Naturwissenschaften* **52**, 15–16.

Maier-Maercker, U. (1979a). 'Peristomatal transpiration' and stomatal movement: A controversial view. I. Additional proof of peristomatal transpiration by hygrophotography and a comprehensive discussion in the light of recent results. *Zeitschrift für Pflanzenphysiologie* **91**, 25–43.

Maier-Maercker, U. (1979b). 'Peristomatal transpiration' and stomatal movement: A controversial view. II. Observation of stomatal movements under different conditions of water supply and demand. *Zeitschrift für Pflanzenphysiologie* **91**, 157–72.

Meidner, H. & Edwards, M. (1975). Direct measurements of turgor pressure potentials of guard cells. I. *Journal of Experimental Botany* **26**, 319–30.

Meidner, H. & Mansfield, T. A. (1968). *Physiology of Stomata.* 179 pp. London: McGraw-Hill.

Osmond, C. B., Ludlow, M. M., Davis, R., Cowan, I. R., Powles, S. B. & Winter, K. (1979). Stomatal responses to humidity in *Opuntia inermis* in relation to control of CO_2 and H_2O exchange patterns. *Oecologia* **41**, 65–76.

Raschke, K. (1976). How stomata resolve the dilemma of opposing priorities. *Philosophical Transactions of the Royal Society of London* Series B, **273**, 551–60.

Raschke, K. (1979). Movements of stomata. In *Encyclopedia of Plant Physiology, New Series*, Vol. 7, ed. W. Haupt & M. E. Feinleib, pp. 383–441. Berlin, Heidelberg, New York: Springer-Verlag.

Raschke, K. & Kühl, U. (1969). Stomatal responses to changes in atmospheric humidity and water supply: Experiments with leaf sections of *Zea mays* in CO_2-free air. *Planta* **87**, 36–43.

Rich, S. & Turner, N. C. (1972). Importance of moisture on stomatal behaviour of plants subjected to ozone. *Journal of the Air Pollution Control Association* **22**, 718–21.

Running, S. W. (1976). Environmental control of leaf water conductance in conifers. *Canadian Journal of Forest Research* **6**, 104–12.

Schulze, E. D. & Küppers, M. (1979). Short-term and long-term effects of plant water deficits on stomatal response to humidity in *Corylus avellana* L. *Planta* **146**, 319–26.

Schulze, E. D., Lange, O. L., Buschbom, U., Kappen, L., Evenari, M. (1972). Stomatal responses to changes in humidity in plants growing in the desert. *Planta* **108**, 259–70.

Schulze, E. D., Lange, O. L., Evenari, M., Kappen, L., Buschbom, U. (1974). The role of air humidity and leaf temperature in controlling stomatal resistance of *Prunus armeniaca* L. under desert conditions. I. A simulation of the daily course of stomatal resistance. *Oecologia* **17**, 159–70.

Schulze, E. D., Lange, O. L., Evenari, M., Kappen, L., Buschbom, U. (1975*a*). The role of air humidity and temperature in controlling stomatal resistance of *Prunus armeniaca* L. under desert conditions, III. The effect on water use efficiency. *Oecologia* **19**, 303–14.

Schulze, E. D., Lange, O. L., Kappen, L., Evenari, M., Buschbom, U. (1975*b*). The role of air humidity and leaf temperature in controlling stomatal resistance of *Prunus armeniaca* L. under desert conditions. II. The significance of leaf water status and internal carbon dioxide concentration. *Oecologia* **18**, 219–33.

Seybold, A. (1961/62). Ergebnisse und Probleme pflanzlicher Transpirationsanalysen. *Jahreshefte der Heidelberger Akademie der Wissenschaften* **6**, 5–8.

Sheriff, D. W. (1977*a*). Where is humidity sensed when stomata respond to it directly? *Annals of Botany* **41**, 1083–4.

Sheriff, D. W. (1977*b*). The effect of humidity on water uptake by, and viscous flow resistance of, excised leaves of a number of species: Physiological and anatomical observations. *Journal of Experimental*

Botany **28**, 1399–407.

Sheriff, D. W. (1979). Stomatal aperture and the sensing of the environment by guard cells. *Plant, Cell and Environment* **2**, 15–22.

Sheriff, D. W. & Kaye, P. E. (1977). Responses of diffusive conductance to humidity in a drought avoiding and a drought resistant (in terms of stomatal response) legume. *Annals of Botany* **41**, 653–5.

Smith, W. K. & Nobel, P. S. (1977). Temperature and water relations for sun and shade leaves of a desert broadleaf, *Hyptis emoryi*. *Journal of Experimental Botany* **28**, 169–83.

Stålfelt, M. C. (1956). Die stomatäre Transpiration und die Physiologie der Spaltöffnungen. *Handbuch der Pflanzenphysiologie*, Vol. 3, ed. O. Stocker, pp. 351–426, Berlin, Heidelberg, New York: Springer-Verlag.

Tenhunen J. D. & Gates, D. M. (1975). Light intensity and leaf temperature as determining factors in diffusion resistance. In *Perspectives of Biophysical Ecology*, ed. D. M. Gates & R. B. Schmerl, *Ecological Studies* **12**, 213–25.

Turner, N. C. & Begg, J. (1973). Stomatal behaviour and water status of maize, sorghum, and tobacco under field conditions. I. At high soil water potential. *Plant Physiology* **51**, 31–6.

Urbach, W., Rupp, W. & Sturm, H. (1976). *Experimente zur Stoffwechselphysiologie der Pflanzen*, 330 pp., Stuttgart: Thieme.

Weatherley, P. E. (1976). Introduction: water movement through plants. *Philosophical Transactions of the Royal Society of London* Series B, **273**, 435–44.

Wright, S. T. C. & Hiron, R. W. P. (1969). (+)-abscisic acid, the growth inhibitor induced in detached wheat leaves by period of wilting. *Nature* **224**, 719–20.

Yemm, E. W. & Willis, A. J. (1954). Stomatal movements and changes of carbohydrate in leaves of *Chrysanthemum maximum*. *New Phytologist* **53**, 373–96.

Ziegler, H. (1967). Wasserumsatz und Stoffbewegungen. *Fortschritte der Botanik* **29**, 68–80.

W. J. DAVIES, J. A. WILSON, R. E. SHARP and
O. OSONUBI

Control of stomatal behaviour in water-stressed plants

Introduction

Plant water deficits occur when the rate of water loss from the plant exceeds the rate of water uptake from the soil. This is a situation which occurs daily, as a result of the contribution of the capacitance of the plant to the transpiration stream. More severe water deficits will arise from a shortage of soil water, either as a result of drying or freezing. Larger deficits will also occur if transpiration rates are very high, for example, when the air is hot or dry. The latter condition may develop even when plants are growing in wet soil. Since the movements of guard cells are mediated through changes in turgor within the epidermis, many stomatal responses to environmental perturbation are modified by the pervasive influence of plant water deficits. Of all the stomatal responses that are considered in this volume, the significance of the stomatal response to substantial leaf water deficit is probably the most easily appreciated.

Small reductions in epidermal turgor, which are routinely experienced by well-watered plants subjected to moderate evaporative demand, can result in increases in leaf conductance. Glinka (1971) has argued that maximum conductances are recorded at optimum water deficits and in more turgid plants, epidermal tissue pressure (superoptimal turgor) may act to reduce leaf conductance. If water is withheld from a plant for several days, water potential will fall to a low value. Under these conditions stomata may remain closed during the light period, since hydraulic considerations will override the photoactive responses of the guard cells. At intermediate levels of stress, it appears that the degree of water deficit influences the stomatal response to photon flux density, temperature, carbon dioxide and vapour pressure deficit to establish a ceiling for stomatal opening at a reduced aperture. Increasing levels of stress will result in greater limitation of the maximum degree of stomatal opening, often accompanied by successively earlier closure of stomata from the maximum aperture for the day (Fig. 1). When plants are rewatered after a

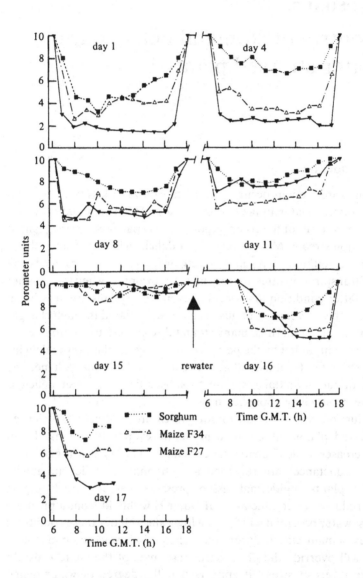

Fig. 1. The effect of increasing plant water stress on the stomatal behaviour (determined by a viscous flow porometer) of three maize and sorghum cultivars. Plants not watered after day 1 and re-watered after day 16.

period of stress it is common to observe a delay of at least some hours, and in some cases days, before stomata reopen (Heath & Mansfield, 1962; Fischer, 1970; Allaway & Mansfield, 1970). This delay occurs despite the recovery of turgor in the plant (Fig. 1).

The stomatal response to decreasing water potential and turgor

There is considerable evidence that at high irradiance and at moderate temperatures and vapour pressure deficits, stomata of many plants remain open until a critical leaf water potential is reached. Thereafter, small increases in water deficit cause marked stomatal closure (Hall, Schulze & Lange, 1976). Several studies (Gardner, 1973; McCree, 1974; van den Driessche, Connor & Tunstal, 1971) have suggested, however, that the leaf conductance of some species is closely coupled to leaf water status so that only very small increases in leaf water deficit will result in some decrease in conductance. Leaf conductance is, however, ultimately determined by turgor levels in the epidermis (Kaufmann, 1976; Lawlor & Lake, 1976; Edwards & Meidner, 1978) and may often be only loosely related to the water potential of the mesophyll cells. Species differences in the shape of the relationship between conductance and bulk leaf water potential may therefore be a function of the degree of hydraulic connection between the mesophyll and the epidermis. In addition, some plants have the capacity to maintain turgor in the guard cells, even though the turgor of the epidermal cells may be decreasing (Brown, Jordan & Thomas, 1976). It seems likely that photon flux density, temperature and vapour pressure deficit during leaf drying, probably as a result of an influence on the rate of leaf drying, will be at least as important as are species differences in determining the shape of the leaf conductance: leaf water potential relationship. Despite the obvious importance of the stomatal response to severe water deficit, this response has generally been considered only in rather imprecise terms. It should be possible to make measurements of mesophyll, guard cell and subsidiary cell turgor of some species. Such measurements would be useful as they would show when and how the guard cells lose turgidity as the leaf dries (Ehret & Boyer, 1979).

The sensitivity of stomata to decreases in leaf water potential varies between species and is influenced by age and growth conditions of plants (Jordan & Ritchie, 1971; McCree, 1974; Davies, 1977). Millar, Gardner & Goltz (1971), Turner & Begg (1973) and Turner (1974) have shown that when stomatal response is related to turgor rather than to water potential, there is a smaller difference between the apparent sensitivities to stress of stomata of different species (Fig. 2). It has long been known that plants from different habitats can exhibit very different solute potentials and it is now apparent that plants may differ markedly in the capacity to accumulate solute as water potential falls. The effect of solute accumulation will be to maintain turgor at even lower water potentials. In the light of these differences, it seems reasonable to suggest that stomata may act in conjunction with solute regulation to maintain a positive turgor in the leaf, rather than acting independently to maintain an apparently arbitrary level of water potential (Cowan, 1977; Davies, Mansfield & Wellburn, 1979). This hypothesis implies the existence of a linkage between bulk leaf turgor and the motive force for stomatal functioning, namely the osmotic potential of the guard cell sap, and identification of the nature of this linkage is clearly of some importance.

The linkage between bulk leaf water status and stomatal behaviour

Stomatal control of leaf turgor occurs *via* a response to a variety of internal and external factors. These responses either initiate metabolic changes that lead to changes in the turgor relations between epidermal and guard cells or promote non-metabolically-mediated changes in epidermal turgor relations. Because of the relative independence of the water relations

Fig. 2. Leaf conductances of several species, at similar irradiances, as a function of (a), leaf water potential, (b), turgor pressure in leaf tissue. Redrawn from Turner (1974). 'Bean' = *Phaseolus vulgaris* L.

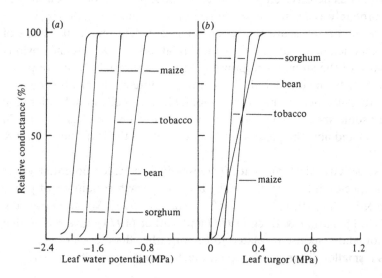

of the epidermis from those of the mesophyll (Edwards & Meidner, 1978) it is tempting to invoke the concept of a chemical messenger, which is produced in the stress-sensitive mesophyll cells when a critical water deficit is reached, moving to the epidermis, where stomatal movements can provide effective control of bulk leaf turgor.

Early evidence for the existence of a chemical 'distress signal' was provided by Glover (1959) who demonstrated that plants which had experienced a period of wilting exhibited delayed reopening of stomata despite the recovery of bulk leaf turgor. In 1969, Wright and his co-workers found that wheat leaves which had been allowed to wilt contained greatly increased amounts of the growth regulator abscisic acid (ABA) (Wright, 1969; Wright & Hiron, 1969). Subsequently, Wright (1977) and others established that the ABA content of leaves of a number of species started to increase at a critical water deficit and from this point the amount of ABA produced by the plant was related to leaf water potential (Fig. 3). Following Wright's early work, Jones & Mansfield (1970, 1972) and others demonstrated that an external application of ABA to leaves resulted in closure of stomata. Stomatal behaviour in plants treated with ABA resembled that exhibited by plants which experienced soil-drying treatment. The case for the involvement of ABA in the processes acting to maintain turgor in the mesophyll

Fig. 3. The effect of leaf water potential on the abscisic acid level in excised wheat leaves. After Wright (1977).

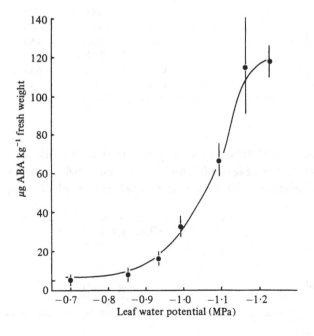

cells is strengthened by the observations of Milborrow & Robinson (1973) and Milborrow (1974*a*) that the chloroplasts are the main sites of ABA formation in green leaves. In addition, Loveys (1977) has shown that the bulk of the stress-stimulated ABA production occurs in the mesophyll and ABA is transported from there to the guard cells (Table 1).

Evidence for the involvement of ABA in stomatal responses to changes in the water status of the plant

Correlations between stomatal behaviour and changes in abscisic acid content

In many studies, closure of stomata as water potential falls may be adequately explained by variation in the ABA content of leaves. Nevertheless, it is clear that under certain circumstances, stress-induced closure of stomata can take place at higher leaf water potentials than those which

Table 1*a*. *ABA generation by stressed and non-stressed* Vicia faba L. *leaf tissue in vitro. Leaf tissue was stressed by exposure to 800 m*M *mannitol for 4.5h before extraction. After Loveys (1977)*

Tissue	ABA, ng/g dry weight		
	Medium	Tissue	Total
Intact non-stressed	236	344	580
+mannitol	740	1350	2090
Mesophyll non-stressed	54	100	154
+mannitol	405	177	582
Epidermis non-stressed	341	313	654
+mannitol	350	200	550

Table 1*b*. *ABA content of* V. faba *leaf tissue from stressed (conditions as above) and non-stressed plants. Figures in brackets represent ABA remaining in epidermis after subtraction of ABA present in adhering mesophyll. After Loveys (1977)*

Tissue	ABA, ng/g dry weight
Intact non-stressed	616
stressed	2608
Epidermis non-stressed	964 (800)
stressed	3812 (2500)

stimulate significant increases in the levels of ABA (e.g. Fig. 4 from Beardsell & Cohen, 1975). Observations of this type have prompted several groups to search for naturally occurring compounds which are active stomatal-closing agents, variations in the level of which might account for stomatal closure which precedes stress-induced increases in ABA levels. Several possible candidates as endogenous regulators of transpiration have been identified (Loveys & Kriedemann, 1974; Ogunkanmi, Wellburn & Mansfield, 1974). In *Sorghum*, the accumulation of one such compound, all-*trans* farnesol, as water potential falls, is closely correlated with stomatal closure (Ogunkanmi *et al.*, 1974). Farnesol will cause reversible closure of stomata when applied to intact leaves (Fenton, Davies & Mansfield, 1977), but its destructive action in certain circumstances (Fenton, Mansfield & Wellburn, 1976) suggests that while the build-up of farnesol in the leaf may be implicated in the stomatal response to water stress, it may not function as a separate endogenous antitranspirant (Mansfield, Wellburn & Moreira, 1978).

Some important measurements of the distribution of ABA within the leaf have provided an alternative explanation for the observation that water-stress-induced stomatal closure can precede a massive build-up of ABA in the plant. Loveys (1977) has shown that nearly all the ABA in the leaves of

Fig. 4. Changes in leaf water potential, leaf resistance and abscisic acid content during a stress–rewatering experiment with maize. All measurements were made 2 h after the start of the photoperiod, except those immediately following rewatering, which were made 4 h after the start of the photoperiod. All resistance measurements in excess of 20 $s\,cm^{-1}$ are shown as $50\,s\,cm^{-1}$. After Beardsell & Cohen (1975).

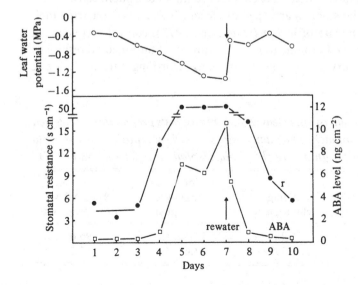

well-watered spinach plants is contained in the chloroplasts (Table 2), and the result of exposing the roots of these plants to 500 mM mannitol for only 4 h, is a release and redistribution of ABA throughout the leaf. It seems clear that if an increase in membrane permeability to ABA occurred at a relatively high water potential then a rapid closure of stomata and extremely fine control of leaf turgor could result simply from a *redistribution* of ABA, without the necessity for any new synthesis. One would postulate, however, that synthesis of new ABA in the chloroplasts would be promoted by its release into the cytoplasm.

Several groups (Wellburn & Hampp, 1976; Milborrow, 1979; Gimmler, Hartung & Heilmann, 1979) have reported on the penetrability of plastid envelopes to ABA and have discussed the possibility that changes in membrane permeability could be used in a regulatory manner. Recent work with ^{14}C ABA by Weyers & Hillman (1979a, b) has shown that stomatal closure occurs when only a very small amount of ABA has reached the guard cells. Small changes of this nature would be effectively masked by relatively large amounts of ABA in the mesophyll. If comparably small quantities of endogenous ABA can initiate stomatal closure, it is possible that they would not be detected by the analytical techniques which are in common use. Indeed, Pierce & Raschke (1979) and Dörffling (1979) have reported stress-stimulated closure of stomata which precedes accumulation of measurable amounts of ABA in the epidermis.

The sensitivity of the guard cells to very low concentrations of ABA may explain the well-documented after-effects of water stress on stomata (Heath & Mansfield, 1962; Fischer, 1970; Allaway & Mansfield, 1970). Weyers & Hillman (1979a) have noted that ABA is accumulated in guard cells regardless of whether the stomata are open or closed. If ABA is compartmentalized away from its site of action (Cummins, 1973), controlled release of small amounts of the hormone following recovery of turgor could result in a delay in the recovery of full stomatal opening. Dörffling, Streich, Kruse &

Table 2. *ABA content of intact leaves and chloroplasts prepared from stressed and non-stressed spinach. The plants were stressed by exposing their roots to 500 mM mannitol for 4h before chloroplast isolation. After Loveys (1977)*

Treatment	Leaf ng/mg chlorophyll	Chloroplasts ng/mg chlorophyll	% of ABA in chloroplasts
Non-stressed	14.9 ± 1.1	14.4 ± 1.1	96.6
Stressed	161.5 ± 9.6	24.6 ± 3.2	15.2

Muxfeldt (1977) have noted that at least in the short term, changes in the ABA content of leaves following rewatering may explain the observed patterns of stomatal behaviour (Fig. 5a, b).

Many field-grown plants exhibit significant diurnal variation in levels of endogenous ABA (Durley, Kannagara & Simpson, 1979). These changes are linked to variation in plant water balance but are not necessarily accompanied by changes in stomatal conductance. Such observations are difficult

Fig. 5. The effect of ABA applied to wilted leaves of *Pisum sativum* L. at time zero for 1 h, (*a*) on the level of ABA in the leaves and (*b*) on the opening of the stomata. Control: wilted leaves treated with water at time zero. A, ABA 10^{-4}M; B, ABA 5×10^{-5}M; C, ABA 2×10^{-5}M. After Dörffling *et al* (1977).

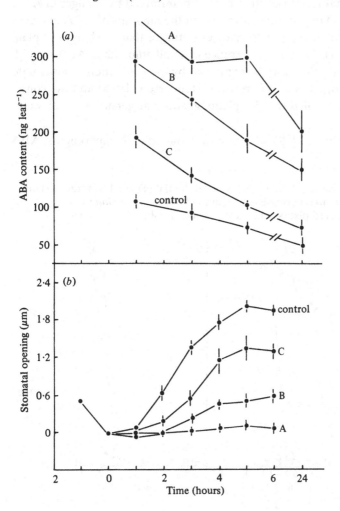

to reconcile with our hypothesis that only small changes in the ABA content of guard cells may be controlling guard cell turgor. One explanation for a lack of stomatal response to quite large changes in ABA may be the apparent insensitivity of stomata of previously stressed plants to subsequent applications or increased endogenous concentration of the hormone (Dörffling *et al.*, 1977). It is important that we should discover more about the hormonal relations of field-grown plants which suffer repeated episodes of stress but generally seem to contain smaller amounts of ABA than laboratory-grown material (Kannangara, Durley & Simpson, 1979).

We have postulated above that stomata may act in conjunction with solute regulation to maintain a minimum leaf turgor, rather than acting independently to maintain an apparently arbitrary leaf water potential. If this is the case, and stomatal responses to stress are determined by changes in ABA content, we would expect that variations in the amounts of ABA in leaves would be linked more closely to changes in turgor than to changes in plant water potential. The general occurrence of small amounts of ABA in field-grown plants might therefore be explained by the generation of low solute potentials (Hsiao, Acevedo, Fereres & Henderson, 1976) and the consequent maintenance of turgor by plants growing at generally lower water potentials.

Pierce & Raschke (1978) have reported that, in their experiments, ABA

Fig. 6. The influence of leaf water potential (*a*) and leaf turgor (*b*) on the accumulation of ABA by leaves of *Capsicum annuum* L. Well-watered plants, triangles; previously wilted plants, circles.

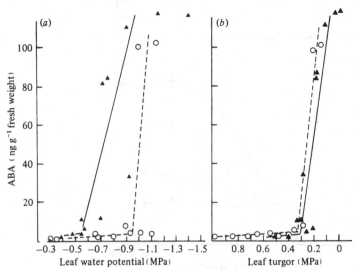

production in leaves appeared to be linked to variation in turgor rather than to variation in water potential. Our own preliminary experiments would support this view. Two groups of *C. annuum* plants, one of which had been previously wilted to promote solute accumulation in leaves, were subjected to a soil drying cycle during which water and solute potentials, ABA levels and stomatal behaviour were measured. The effect of pre-stressing was to lower the water potential threshold for increased ABA production (Fig. 6*a*) and for stomatal closure. Despite this, ABA production (Fig. 6*b*) and stomatal closure were initiated in the two groups of plants, at approximately the same threshold turgor.

The extent to which external application of ABA can substitute for the effect of water stress on stomata

Mansfield (1976*a*) noted that spraying low concentrations of ABA on to leaves of a variety of plants induces a pattern of stomatal behaviour not unlike that produced by withholding water from the plant (Fig. 1). It is well known that withholding water has a marked influence on the stomatal response to many environmental variables. For example, Heath & Mansfield (1962) noted that the stomatal closing response when plants were exposed to high concentrations of carbon dioxide was often enhanced if the plants remained unwatered for a few days prior to treatment. Wilson (1980) has confirmed this result by subjecting epidermal strips to variations in carbon dioxide concentration (Table 3). Strips were removed either from well-watered plants or from plants which had not been watered for two days (decrease in water potential of 0.3 MPa). Interestingly, stomata in strips removed from plants previously sprayed with 10^{-4}M ABA showed a similar response to that shown by stomata of water-stressed plants. Although ABA apparently completely substituted for the effects of water stress, it should be noted that the incubation of stomata from water-stressed plants on an aqueous medium is a rather artificial situation.

With some exceptions (Squire, 1979; Black & Squire, 1979), stomatal

Table 3. *The effect of carbon dioxide on stomatal aperture (μm) of well-watered, water-stressed and ABA treated plants. Water-stressed plants remained unwatered for two days (reduction in water potential of -0.3 MPa) and ABA-treated plants were sprayed to run-off with a 10^4M aqueous solution of $\pm$ ABA. After Wilson (1980)*

	Well-watered	Water-stressed	ABA treated
$-CO_2$	18.5 ± 0.3	17.1 ± 0.3	17.1 ± 0.3
$+CO_2$	13.2 ± 0.3	7.7 ± 0.2	6.7 ± 0.2

responses to humidity are enhanced when plants are mildly stressed (Schulze *et al.*, 1973; Osonubi & Davies, 1980) (Fig, 7). It appears, however, that the humidity response in plants treated with ABA is of a similar magnitude to that exhibited by well-watered plants (P. J. Ferrar, O. Osonubi and W. J. Davies, unpublished data). Stomatal closure at high vapour pressure deficit is caused by vapour loss directly from the stomatal apparatus. One conclusion from these data is, therefore, that the effect of water stress over and above the effect promoted by ABA is either to increase the resistance to water movement to the stomata or to decrease the resistance to water loss from the guard cell apparatus. The effect of water stress on stomata may therefore result both from a direct influence of ABA on guard cell turgor and from an effect of the water deficit on the hydraulic conductivity of the leaf tissues.

A role for ABA in drought endurance, drought avoidance and optimization of gas exchange?

There are several processes, other than stomatal closure, which will contribute to the maintenance of plant turgor during times of water shortage. Of particular significance in this regard is the observation that the application of ABA to root tissue can result in a stimulation of volume flow of water and ion flux across the root (Glinka & Reinhold, 1971; Collins & Kerrigan, 1973; Karmoker & Van Steveninck, 1978). It is tempting to

Fig. 7. The effect of ABA or water stress (WS) on the stomatal response to humidity in *Betula pendula* Roth. WW, well watered controls. (P. J. Ferrar, O. Osonubi & W. J. Davies, unpublished data.)

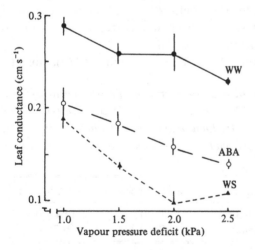

suggest that ABA might act to maintain turgor by decreasing transpiration and simultaneously promoting an increased flux of water into the roots. It is known that ABA can move through the phloem to the roots from water-stressed leaves (Hocking, Hillman & Wilkins, 1972; Hoad, 1973) but it remains to be shown that this ABA influences water uptake by the stressed plant.

It is well-known that in many systems ABA has an inhibitory effect upon growth (Milborrow, 1974*b*). Davies *et al.* (1979) have suggested that the limitation of leaf extension, which is often recorded when water is withheld from plants (Fig. 8*a*), may be caused by a build-up of ABA in the leaf. Clearly, a reduction in the rate of leaf area development is in itself an advantage to a plant which is suffering from reduced water availability. In addition to this, the increased availability of assimilates as a result of reduced leaf growth can result in an accumulation of osmotically active substances in leaves and roots, with the possible consequence that turgor may be maintained in plants growing in drying soil. The phenomenon of osmoregulation might be particularly effective if ABA reduced the rate of cell expansion and the ultimate size of leaf cells (Quarrie & Jones, 1977).

Rather than inhibiting the growth of roots, soil drying can stimulate root growth and increase root : shoot ratio (Hsiao & Acevedo, 1974; Sharp & Davies, 1979). Interestingly, spraying low concentrations of ABA on to leaves can also cause an increase in root development and may even result in a net increase in root growth compared to root growth by well-watered plants (Table 4, Watts, Rodriguez & Davies, 1980). An increase in root growth may have some advantage in increasing the amount of water available to a plant growing in drying soil.

Sharp & Davies (1979) and Davies *et al.* (1979) have shown that stress-induced limitation in leaf growth, stimulation of root growth and solute accumulation are often accompanied by only a very small degree of stomatal closure (Fig. 8*b*), and followed only at larger water deficits by complete closure of stomata. A combination of these turgor-maintaining events rather than rapid complete closure of stomata would enable photosynthesis to continue at a relatively high rate for an extended period. A continuous supply of photosynthates is important for continued solute regulation and continued growth, especially of the roots. It seems likely that some turgor-maintaining processes might be promoted by the *redistribution* of ABA in the leaf at a comparatively high water potential. It might be argued, therefore that in this situation, ABA provides the plant with some drought-enduring capacity. While turgor is maintained in the leaf, ABA production is maintained at a low level and only later in the drying cycle, when water potential and leaf turgor have dropped to a low and potentially damaging

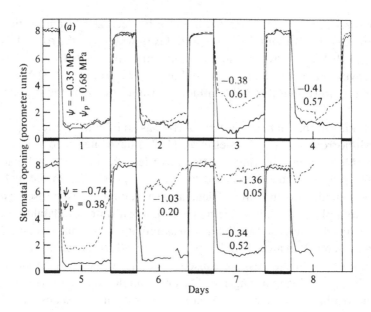

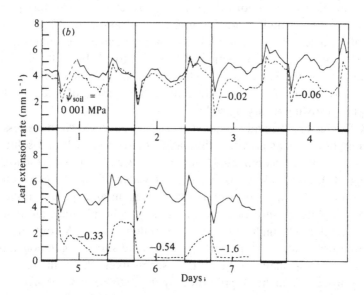

Fig. 8. The influence of developing leaf water stress, (a) on stomatal behaviour and, (b) on leaf growth by maize seedlings (interrupted lines, water-stressed plants, full lines, well-watered plants). Leaf and soil water potentials and turgor levels are also shown. After Davies et al (1979).

Table 4. *The influence of water stress and ABA on root and shoot growth of Capsicum. After Watts et al. (1980). Water was withheld from water-stressed plants for successive periods of 7 days prior to re-watering. ABA applied 13, 19 and 25/6/79.*

Treatment	Harvest 5/6/79 Shoot:root ratio	Harvest 12/6/79 Shoot:root ratio	Harvest 19/6/79 Shoot:root ratio	Harvest 26/6/79 Shoot:root ratio
Well-watered	7.043 ± 0.491	7.226 ± 0.390	7.216 ± 0.431	7.270 ± 0.271
Well-watered plus ABA	7.816 ± 0.892	8.289 ± 1.253	6.898 ± 0.474	5.576 ± 0.444
Water-stressed	6.674 ± 0.672	7.047 ± 0.562	6.250 ± 0.345	5.625 ± 0.278
Water-stressed plus ABA	6.704 ± 0.685	7.215 ± 0.261	7.376 ± 0.445	6.120 ± 0.558

level, will abnormal ABA synthesis occur. This synthesis will promote almost complete stomatal closure and therefore a stress-avoiding state.

Several pieces of evidence have suggested that there may be some inter-relation between the responses of stomata to carbon dioxide and to ABA. Raschke (1975) has found that the stomata of some species have a low sensitivity to carbon dioxide when leaves have a high turgor. He argued that this is because the ABA levels in the guard cells are low. Mansfield (1976b) and Mansfield & Wilson (1979) have presented evidence which suggests that the effects of ABA and carbon dioxide applied together are essentially additive. The results of our own experiments (Table 3) show that water stress or ABA treatment *prior* to an exposure to high carbon dioxide levels will increase the sensitivity of guard cells to carbon dioxide. Experiments of this type suggest that it is unlikely that an adequate picture of stomatal responses to ABA and/or carbon dioxide will be obtained if carbon dioxide or single applications of hormones are applied only to well-watered plants (Davies, 1978; Mansfield & Wilson, 1979). Clearly, the question of whether ABA interacts with carbon dioxide or not is of considerable significance in our understanding of the control of stomata in water-stressed plants. Dubbe, Farquhar & Raschke (1978) have concluded that the influence of ABA on the gain of a feedback loop involving intercellular carbon dioxide and stomata may enable the stomata to optimize gas exchange when leaf turgor is reduced. In contrast, Mansfield & Wilson (1979) have cited the after-effect of water stress on stomata as one example of a situation in which stomata are exhibiting rather crude and arbitrary control in relation to external conditions rather than the proportional control proposed by Dubbe *et al.* (1978).

Further investigations of the factors influencing the distribution and effects of ABA

In the light of the evidence which we have reviewed above, it is arguable that the ability to form ABA has been looked upon as a desirable character only in rather imprecise terms (Mansfield & Wilson, 1979). We have proposed that the level of ABA in the guard cell is critical in determining the stomatal response to developing stress. Therefore, it seems reasonable that future research should be directed at determining changes in ABA levels in different plant parts in response to various treatments, and defining the mechanisms which are responsible for the redistribution of ABA throughout the plant.

In our recent work we have noted that as well as stimulating changes in ABA levels, increases in plant water deficit can result in large changes in the endogenous levels of all-*trans* farnesol (Fig. 9) (Ogunkanmi *et al.*, 1974; Wellburn, Ogunkanmi, Fenton & Mansfield, 1974; Wilson & Davies, 1979) and massive changes in the levels of free fatty acids (Table 5). Fenton *et al.* (1976) have shown that both of these are capable of inhibiting the metabolic functions of isolated chloroplasts probably as a result of the effects of the compounds upon envelope membranes. Mansfield *et al.* (1978) and Mansfield & Wilson (1979) have proposed that the stress-stimulated build-up of either or both of these membrane-active compounds may be related to the stress-stimulated release of ABA from mesophyll chloroplasts. It is

Fig. 9. The effects of developing water stress on stomatal behaviour and levels of farnesol-like antitranspirants in *Zea mays* L. cv. Farz 27. Plants not watered after day 0 and rewatered after day 14. Redrawn from Wilson & Davies (1979).

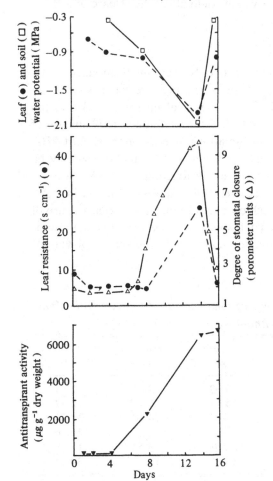

well-known that the application of both farnesol and fatty acids to intact leaves, isolated epidermis or to chloroplasts can cause ultrastructural effects (Siegenthaler, 1969; Fenton *et al.*, 1976), not unlike those caused by the development of severe water stresses (Vieira-da-Silva, Naylor & Kramer, 1974; Freeman & Duysen, 1975; Giles, Beardsell & Cohen, 1974). It seems likely that low concentrations of the membrane active compounds which are associated with water stress in many plants could act in a more subtle manner than the high concentrations found in severely stressed tissue. Information is therefore needed on the effects of moderate levels of water stress on the biosynthesis and metabolism of lipids, the structure of membranes and on the various stages of the terpenoid biosynthetic pathway. The increase in linolenic acid, which we have found to be associated with water stress in maize and *Sorghum* (Table 5), might point to an increase in the unsaturation of membranes leading to changes which could affect the release of ABA.

Conclusions

It is now clear that the presence of only a very small amount of ABA at a given site of action (e.g. the root tip or the guard cell) may stimulate a significant response. An elucidation of the involvement of ABA in the responses of the plant to water deficits therefore requires careful analysis of ABA distribution patterns. More information is needed on the sites of ABA synthesis, in particular whether the development of mild stress may be detected in the roots via the increased production of ABA (Walton, Harrison & Cote, 1976). The processes and pathways involved in ABA movement, particularly from the chloroplasts to the guard cells, are also of crucial importance since the speed of response to a reduction of leaf turgor is of some significance. In addition, the comparative sensitivity of various

Table 5. *Free acids as a percentage of total acids in* Zea mays *(John Innes hybrid) after different periods without water*

| Fatty Acid | Days without water | | | |
	1	8	19	21
<C–16	22.40	25.62	19.48	22.63
16:0	10.44	14.08	9.40	18.57
18:0	–	–	14.10	32.71
18:1	–	–	–	26.56
18:2	–	5.58	5.53	17.78
18:3	5.51	13.39	12.68	39.14

processes to ABA, as well as the possible effects of preconditioning should be determined, since field-grown plants clearly do not remain 'well-watered' throughout their growth period.

Many of the arguments advanced in the present paper have led Mansfield & Wilson (1979) to suggest that a search for plants exhibiting *rapid* release of ABA to the sites of action either in the roots or in the leaves may provide an advance in our appreciation of the processes contributing to drought tolerance. We would suggest that a study of the turgor relations and ABA distribution throughout plants exhibiting only small water deficits may be of greater direct relevance in such studies than considerations of the amounts of ABA produced under severe water stress. Other results would suggest that effects of stress over and above those possibly mediated through ABA may have an important influence on plant water balance.

References

Allaway, W. G. & Mansfield, T. A. (1970). Experiments and observations on the after-effect of wilting on stomata of *Rumex sanguineus. Canadian Journal of Botany* **48**, 513–21.

Beardsell, M. F. & Cohen, D. (1975). Relationships between leaf water status, abscisic acid levels and stomatal resistance in maize and sorghum. *Plant Physiology* **56**, 207–12.

Black, C. R. & Squire, G. R. (1979). Effects of atmospheric saturation deficit on the stomatal conductance of pearl millet and groundnut. *Journal of Experimental Botany* **30**, 935–46.

Brown, K. W., Jordan, W. R. & Thomas, J. C. (1976). Water stress induced alterations of the stomatal response to decreases in leaf water potential. *Physiologia Plantarum* **37**, 1–5.

Collins, J. C. & Kerrigan, A. P. (1973). Hormonal control of ion movements in the plant root. In *Ion transport in plants*, ed. W. P. Anderson, pp. 589–94. London: Academic Press.

Cowan, I. R. (1977). Stomatal behaviour and environment. In *Advances in Botanical Research* Vol. IV, ed. R. D. Preston & H. W. Woolhouse, pp. 117–223. London: Academic Press.

Cummins, W. R. (1973). The metabolism of abscisic acid in relation to its reversible action on stomata in leaves of *Hordeum vulgare* L. *Planta* **114**, 159–67.

Davies, W. J. (1977). Stomatal responses to water stress and light in plants grown in controlled environments and in the field. *Crop Science* **17**, 735–40.

Davies, W. J. (1978). Some effects of abscisic acid and water stress on stomata of *Vicia faba* L. *Journal of Experimental Botany* **29**, 175–82.

Davies, W. J., Mansfield, T. A. & Wellburn, A. R. (1979). A role for abscisic acid in drought endurance and drought avoidance. In *Proceedings of the Tenth International Conference on Plant Growth Substances*, ed. F. Skoog (in the press).

Dörffling, K. H. (1979). Studies on the physiologic role of abscisic acid. *Abstracts of Tenth International Conference on Plant Growth Substances*, p. 22.

Dörffling, K., Streich, J., Kruse, W. & Muxfeldt, B. (1977). Abscisic acid and the after-effect of water stress on stomatal opening potential. *Zeitschrift für Pflanzenphysiologie* **31**, 43–56.

Driessche, van den R., Connor, D. J. & Tunstal, B. R. (1971). Photosynthetic response of brigalow to irradiance, temperature and water potential. *Photosynthetica* **5**, 210–17.

Dubbe, D. R., Farquhar, G. D. & Raschke, K. (1978). Effect of abscisic acid on the gain of the feedback loop involving carbon dioxide and stomata. *Plant Physiology* **62**, 413–17.

Durley, R. C., Kannangara, T. & Simpson, G. M. (1979). Analysis of abscisins, cytokinins and auxins in *Sorghum bicolor* leaves by high performance liquid chromatography and diurnal fluctuation of these three hormones. *Abstracts of Tenth International Conference on Plant Growth Substances*, p. 22.

Edwards, M. & Meidner, H. (1978). Stomatal responses to humidity and the water potentials of epidermal and mesophyll tissue. *Journal of Experimental Botany* **29**, 711–80.

Ehret, D. L. & Boyer, J. S. (1979). Potassium loss from stomatal guard cells at low water potentials. *Journal of Experimental Botany* **30**, 225–34.

Fenton, R., Davies, W. J. & Mansfield, T. A. (1977). The role of farnesol as a regulator of stomatal opening in *Sorghum*. *Journal of Experimental Botany* **28**, 1043–53.

Fenton, R., Mansfield, T. A. & Wellburn, A. R. (1976). Effects of isoprenoid alcohols on oxygen exchange of isolated chloroplasts in relation to their possible physiological effects on stomata. *Journal of Experimental Botany* **27**, 1206–14.

Fischer, R. A. (1970). After effects of water stress on stomatal opening potential. II. Possible causes. *Journal of Experimental Botany* **21**, 386–404.

Freeman, T. P. & Duysen, M. E. (1975). The effect of imposed water stress on the development and ultrastructure of wheat chloroplasts. *Protoplasma* **83**, 131–45.

Gardner, R. W. (1973). Internal water status and plant response in relation to the external water regime. In *Plant Response to Climatic Factors*. Proceedings of Uppsala Symposium, pp. 221–5, UNESCO.

Giles, K. L., Beardsell, M. F. & Cohen, D. (1974). Cellular and ultrastructural changes in mesophyll and bundle sheath cells of maize in response to water stress. *Plant Physiology* **54**, 208–12.

Gimmler, H., Hartung, W. & Heilmann, B. (1979). ABA-permeability of the chloroplast envelope, its distribution between chloroplasts and cytoplasm, and its effect on photosynthesis. *Abstracts of Tenth International Conference on Plant Growth Substances*, p. 23.

Glinka, Z. (1971). The effect of epidermal cell water potential on stomatal response to illumination of leaf discs of *Vicia faba*. *Physiologia Plantarum* **24**, 476–9.

Glinka, Z. & Reinhold, L. (1971). Abscisic acid raises the permeability of plant cells to water. *Plant Physiology* **48**, 103–5.

Glover, J. (1959). The apparent behaviour of maize and sorghum stomata during and after drought. *Journal of Agricultural Science* **53**, 412–16.

Hall, A. E., Schulze, E.-D. & Lange, O. L. (1976). Current perspectives of steady-state stomatal responses to environment. In *Water and Plant*

Life. Ecological Studies, Analysis and Synthesis. Vol. 19, ed. O. L. Lange *et al.* New York: Springer-Verlag.

Heath, O. V. S. & Mansfield, T. A. (1962). A recording porometer with detachable cups operating on four separate leaves. *Proceedings of the Royal Society of London, Series B* **156**, 1–13.

Hoad, G. V. (1973). Effect of moisture stress on abscisic acid levels in *Ricinus communis* L. with particular reference to phloem exudate. *Planta* **113**, 367–72.

Hocking, T. J., Hillman, J. R. & Wilkins, M. B. (1972). Movement of abscisic acid in *Phaseolus vulgaris* plants. *Nature New Biology* **235**, 124–5.

Hsiao, T. C. & Acevedo, E. (1974). Plant responses to water deficits, water use efficiency and drought resistance. *Agricultural Meteorology* **14**, 59–84.

Hsiao, T. C., Acevedo, E., Fereres, E. & Henderson, D. W. (1976). Water stress, growth and osmotic adjustment. *Philosophical Transactions of the Royal Society of London, Series B* **273**, 479–500.

Jones, R. J. & Mansfield, T. A. (1970). Suppression of stomatal opening in leaves treated with abscisic acid. *Journal of Experimental Botany* **21**, 714–19.

Jones, R. J. & Mansfield, T. A. (1972). Effects of abscisic acid and its esters on stomatal aperture and the transpiration ratio. *Physiologia Plantarum* **26**, 321–7.

Jordan, W. R. & Ritchie, J. T. (1971). Influence of soil water stress on evaporation, root absorption and internal water status of cotton. *Plant Physiology* **48**, 783–8.

Kaufmann, M. R. (1976). Stomatal response of Englemann spruce to humidity, light and water stress. *Plant Physiology* **57**, 898–901.

Kannangara, T., Durley, R. C. & Simpson, G. M. (1979). Changes in hormones during development of sorghum grown under irrigated and stress conditions. *Abstracts of Tenth International Conference on Plant Growth Substances*, p. 22.

Karmoker, J. L. & Van Steveninck, R. F. M. (1978). Stimulation of volume flow and ion flux by abscisic acid in excised root systems of *Phaseolus vulgaris* L. cv. Redland Pioneer. *Planta* **141**, 37–43.

Lawlor, D. W. & Lake, J. V. (1976). Evaporation rate, leaf water potential and stomatal conductance in *Lolium, Trifolium* and *Lysimachia* in drying soil. *Journal of Applied Ecology* **13**, 639–46.

Loveys, B. R. (1977). The intracellular location of abscisic acid in stressed and non-stressed leaf tissue. *Physiologia Plantarum* **40**, 6–10.

Loveys, B. R. & Kriedemann, P. E. (1974). Hormonal regulation of gas exchange. In *Mechanisms of Regulation of Plant Growth, Bulletin 12*, ed. R. L. Bieleski *et al.*, pp. 781–787. The Royal Society of New Zealand.

McCree, K. J. (1974). Changes in the stomatal response characteristics of grain sorghum produced by water stress during growth. *Crop Science* **14**, 273–8.

Mansfield, T. A. (1976a). Chemical control of stomatal movements. *Philosophical Transactions of the Royal Society of London, Series B* **273**, 541–50.

Mansfield, T. A. (1976b). Delay in the response of stomata to abscisic acid in CO_2-free air. *Journal of Experimental Botany* **27**, 559–64.

Mansfield, T. A., Wellburn, A. R. & Moreira, T. J. S. (1978). The role of abscisic acid and farnesol in the alleviation of water stress. *Philosophical Transactions of the Royal Society of London, Series B* **284**, 471–82.

Mansfield, T. A. & Wilson, J. (1979). Regulation of gas exchange in water-stressed plants. *Proceedings of the Nottingham Easter School* (in the press).

Milborrow, B. V. (1974a). Biosynthesis of abscisic acid by a cell-free system. *Phytochemistry* **13**, 131–6.

Milborrow, B. V. (1974b). The chemistry and physiology of abscisic acid. *Annual Review of Plant Physiology* **25**, 259–307.

Milborrow, B. V. (1979). Antitranspirants and the regulation of abscisic acid content. *Australian Journal of Plant Physiology* **6**, 249–54.

Milborrow, B. V. & Robinson, D. R. (1973). Factors affecting the biosynthesis of abscisic acid. *Journal of Experimental Botany* **24**, 537–48.

Millar, A. A., Gardner, W. R. & Goltz, S. M. (1971). Internal water status and water transport in seed onion plants. *Agronomy Journal* **63**, 779–84.

Ogunkanmi, A. B., Wellburn, A. R. & Mansfield, T. A. (1974). Detection and preliminary identification of endogenous antitranspirants in water-stressed *Sorghum* plants. *Plants* **117**, 293–302.

Osonubi, O. & Davies, W. J. (1980). The influence of plant water stress on stomatal control of gas exchange at different levels of atmospheric humidity. *Oecologia* **46**, 1–6.

Pierce, M. & Raschke, K. (1978). The relationship between abscisic acid accumulation and leaf turgor. *Plant Physiology* (Suppl.) **61**, 131.

Pierce, M. & Raschke, K. (1979). Increases in abscisic acid in leaf epidermis during water stress. *Abstracts of the Tenth International Conference on Plant Growth Substances*, p. 49.

Quarrie, S. A. & Jones, H. G. (1977). Effects of abscisic acid and water stress on development and morphology of wheat. *Journal of Experimental Biology* **28**, 192–203.

Raschke, K. (1975). Stomatal action. *Annual Review of Plant Physiology* **26**, 309–40.

Schulze, E.-D., Lange, O. L., Kappen, L., Buschbom, U. & Evanari, I. (1973). Stomatal responses to changes in temperature at increasing water stress. *Planta* **110**, 29–42.

Sharp, R. E. & Davies, W. J. (1979). Solute regulation and growth by roots and shoots of water-stressed maize plants. *Planta* **147**, 43–9.

Siegenthaler, P. A. (1969). Vieillissement de l'appareil photosynthétique. I. Effect synergigne de la lumière et du vieillissement *in vitro* sur les changements de colume de chloroplastes isolés d'épinard. *Plant and Cell Physiology* **10**, 801–10.

Squire, G. R. (1979). The response of stomata of Pearl millet (*Pennisetum typhoides* S. & H.) to atmospheric humidity. *Journal of Experimental Botany* **30**, 925–34.

Turner, N. C. (1974). Stomatal responses to light and water under field conditions. In *Mechanisms of Regulation of Plant Growth, Bulletin 12,* ed. R. Bieleski *et al.*, pp. 423–32. Royal Society of New Zealand.

Turner, N. C. & Begg, J. E. (1973). Stomatal behaviour and water status of maize, sorghum and tobacco under field conditions. *Plant Physiology* **51**, 31–6.

Vieira-da-Silva, J., Naylor, A. W. & Kramer, P. J. (1974). Some ultrastructural and enzymatic effects of water stress in cotton *(Gossypium hirsutum L.)* leaves. *Proceedings of the National Academy of Science of the USA* **71**, 3243–7.

Walton, D. C., Harrison, M. A. & Cote, P. (1976). The effects of water stress on abscisic acid levels and metabolism in roots of *Phaseolus vulgaris* L. and other plants. *Planta* **131**, 141–4.

Watts, S. Rodriguez, J. L. & Davies, W. J. (1980). Root and shoot growth of plants treated with abscisic acid. *Annals of Botany* (in the press).

Wellburn, A. R., Ogunkanmi, A. B., Fenton, R. & Mansfield, T. A. (1974). All-*trans* farnesol: a naturally occurring antitranspirant? *Planta* **120**, 255–63.

Wellburn, A. R. & Hampp, R. (1976). Fluxes of gibberellic and abscisic acids, together with that of adenosine 3',5'-cyclic phosphate, across plastid envelopes during development. *Planta* **131**, 95–6.

Weyers, J. D. B. & Hillman, J. R. (1979*a*). Uptake and distribution of abscisic acid in *Commelina* leaf epidermis. *Planta* **144**, 167–72.

Weyers, J. D. B. & Hillman, J. R. (1979*b*). Sensitivity of *Commelina* stomata to abscisic acid. *Planta* **146**, 623–8.

Wilson, J. A. (1980). Stomatal responses to applied ABA and CO_2 in epidermis detached from well-watered and water-stressed plants of *Commelina communis* L. *Journal of Experimental Botany* (in the press).

Wilson, J. A. & Davies, W. J. (1979). Farnesol-like antitranspirant activity and stomatal behaviour in maize and *Sorghum* lines of differing drought tolerance. *Plant, Cell and Environment* **2**, 49–57.

Wright, S. T. C. (1969). An increase in the 'inhibitor-β' content of detached wheat leaves following a period of wilting. *Planta* **36**, 10–20.

Wright, S. T. C. (1977). The relationships between leaf water potential (ψ leaf) and the levels of abscisic acid and ethylene in excised wheat leaves. *Planta* **134**, 183–9.

Wright, S. T. C. and Hiron, R. W. P. (1969). (+) abscisic acid, the growth inhibitor in detached wheat leaves following a period of wilting. *Nature* **224**, 719–20.

M. H. UNSWORTH & V. J. BLACK

Stomatal responses to pollutants

Introduction

There are two distinct reasons for studying stomatal responses to
pollutant gases. First, stomata are important in regulating the rates of uptake
of gaseous pollutants by leaves. Second, stomatal responses to pollutants
modify the normal exchange processes of water vapour, carbon dioxide and
oxygen between leaves and the atmosphere. Thus stomatal responses may
influence two consequences of exposure to pollutants: *damage* to tissue
either by accumulation of a pollutant or from toxic products generated by
reactions of pollutants within the leaf; physiological or biochemical *stress*
induced for example by modifications in water relations, enzyme activity or
ionic balances.

Several extensive reviews of physiological and biochemical responses of
plants to pollutants have been published recently (Verkroost, 1974; Mudd
& Kozlowski, 1975; Hällgren, 1978). In this chapter we aim to summarize
the evidence of stomatal responses, particularly to sulphur dioxide, ozone
and nitrogen dioxide, singly and as mixtures and to indicate interactions with
other environmental factors. Coverage is not intended to be exhaustive;
rather, examples are chosen to illustrate our current understanding of forms
of stomatal response.

When Mansfield reviewed this topic in 1973 he added a postscript to
emphasize the 'dearth of satisfactory information from well-designed,
critical experiments'. In the intervening years the situation has improved
only marginally. There have been remarkably few experiments designed
specifically to study stomatal responses to pollutants. Too often, sweeping
conclusions concerning stomatal responses are drawn from laboratory
experiments in which transpiration has been measured crudely in ill-defined
conditions. Laboratory experiments are often carried out at very high pollutant concentrations, atypical of those found in the real atmosphere. In
designing exposure cabinets, the importance of adequate air movement to
ensure that gas exchange is not limited by boundary layer resistances

(Šesták, Čatský & Jarvis, 1971) is sometimes insufficiently appreciated, in spite of several studies demonstrating the influence of boundary layer resistance on responses to pollutants (Brennan & Leone, 1968; Ashenden & Mansfield, 1977; Black & Unsworth, 1979a).

In fairness, it must also be emphasized that the study of stomatal responses to pollutants poses a number of experimental problems in addition to those faced by the usual stomatal physiologist. For example, many air pollutants are readily absorbed or adsorbed on to walls of leaf chambers, tubing and leaf cuticle (Spedding, 1969). Consequently knowledge of the flux of pollutant into leaves is usually obtained only by indirect methods of analysis (Black & Unsworth, 1979c; Unsworth, 1981). Gas analysers for some pollutants lack sensitivity, and so encourage a tendency for experiments at high gas concentrations. Since many pollutants appear to influence both photosynthesis and stomatal movement, the distinction between direct stomatal responses to pollutants and those resulting from action on carbon dioxide exchange within the mesophyll is difficult to draw.

In spite of these problems, we believe that for several pollutants the weight of evidence shows consistent forms of response and this evidence will now be outlined.

Measurements of responses

Ozone

Naturally occurring ozone concentrations at sea level seldom exceed 40 ppb (parts in 10^9 by volume). Ozone as a pollutant is generated by the action of sunlight on vehicle exhaust gases and reaches concentrations of 200–300 ppb in appropriate weather at many locations in North America (Vukovich, Bach, Crissman & King, 1977), Europe (Harrison & Holman, 1979) and other regions with high densities of motor vehicles. Some of the first visual damage to vegetation by ozone was detected on sensitive varieties of tobacco (Mukammal, 1965). Tobacco farmers in affected areas found that crops were least sensitive when water stressed, suggesting that stomatal uptake was an important pathway for ozone. Laboratory experiments by Rich, Waggoner & Tomlinson (1970) and Smith and King (in Unsworth, 1980) confirmed this hypothesis, and demonstrated that ozone concentrations at mesophyll cell walls appear to be close to zero. Thus stomatal conductance generally provides the major control of ozone uptake.

Stomata of many species respond to ozone by partially closing. Fig. 1 shows results of Hill & Littlefield (1969) for oat plants (*Avena sativa* L.) exposed to 600 ppb of ozone in a wind tunnel. Net photosynthesis decreased over a period of about 1h on exposure to ozone and recovered partially when ozonation ceased. Stomatal aperture (measured from epidermal strips) was reduced within 20 min by ozone. Similar magnitudes of stomatal closure were found at 100 and 200 ppb ozone, and in all cases reopening was reported after removal of the pollutant.

Hill & Littlefield suggested that the stomatal responses they observed arose independently of any effects of ozone on photosynthesis or respiration, but crucial experiments proposed by Mansfield (1973) to clarify this point do not yet appear to have been reported.

If the stomatal closure as a response to ozone occurs rapidly, does it protect plants from damage by restricting fluxes of ozone into leaves? Rich & Turner (1972) studied stomatal responses to ozone of varieties of tobacco known to be sensitive and resistant to visible damage by ozone in an attempt to see whether stomatal responses could explain the resistance. Provided that atmospheric humidity was low, stomata of the resistant variety responded in two ways which restricted ozone uptake and so conferred protection: (i) stomata closed in response to saturation deficit, irrespective of ozone concentration; (ii) on exposure to ozone, stomata closed further, and much more rapidly than those of the sensitive variety. Several other reports (Adedipe, Khatamian & Ormrod, 1973; Hill & Littlefield, 1969;

Fig. 1. Responses of net photosynthesis (full lines) and stomatal width (interrupted lines) of oats plants exposed either to 600 ppb ozone for 1 h (circles) or in clean air (square symbols). Courtesy of Hill & Littlefield (1969).

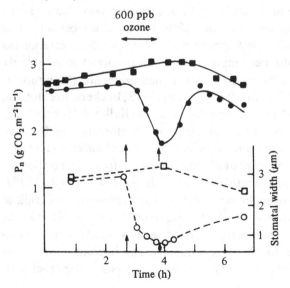

Knudson Butler & Tibbitts, 1979) support the view that stomatal responses may be a factor in determining resistance to ozone damage.

Although, with few exceptions (notably work of Dugger, Ting and co-workers, e.g. Dugger & Ting, 1970; Evans & Ting, 1974) it is agreed that exposure to ozone leads to stomatal closure, the mechanism of response is still unclear. Heath (1975) and Verkroost (1974) recently reviewed the substantial literature concerning biochemical and cellular effects of ozone, but the links between cellular effects and stomatal responses have yet to be established. In principle ozone is such a reactive gas that it might be expected to react with the first structures it encounters, e.g. cell walls or plasmalemma (Heath, 1975). In practice the distribution of sinks for ozone must be much more complex in the intact plant to account for responses such as changes in rates of photosynthesis.

Sulphur dioxide

Sulphur dioxide in the atmosphere arises predominantly from the burning of fossil fuels (especially coal) and the smelting of ores. Daily mean concentrations in urban areas or in industrial regions are generally in the range 10–40 ppb and short-term fluctuations lasting a few minutes may be a factor of five or more times larger (Martin & Barber, 1973).

Some of the first measurements of stomatal responses to SO_2 were made by Mansfield & Majernik (1970; Majernik & Mansfield, 1971) using *Vicia faba* L. exposed to about 500 ppb SO_2. They found that SO_2 induced stomatal opening when atmospheric relative humidity was above about 40% (≈ 1.7 kPa vapour pressure deficit, vpd), and closure occurred at lower humidities. The opening response for *V. faba* in humid atmospheres was confirmed by Unsworth, Biscoe & Pinckney (1972) at SO_2 concentrations as low as 25 ppb, and has also been reported for other species exposed at fairly low SO_2 concentrations both from direct measurements (Majernik & Mansfield, 1970; Biscoe, Unsworth & Pinckney, 1973; Beckerson & Hofstra, 1979 *a,b*) and indirectly by measuring transpiration (Keller & Muller, 1958; Ashenden 1979). In most species the increase in stomatal conductance during exposures up to about three days is about 20% of the clean air value and, surprisingly, is independent of SO_2 concentration for concentrations in the range 18–350 ppb (Unsworth *et al.*, 1972; Black & Unsworth, 1979*a*, 1980). There is debate over the reversibility of the response: Majernik & Mansfield (1970) found that when plants were returned to clean air, conductances returned to control values if plants had been exposed for up to 6 h to SO_2, but with exposures for three days the increases in conductance were irreversible. Unsworth *et al.* (1972) also found reversible responses with

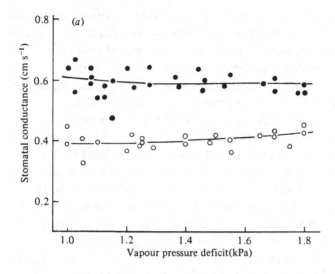

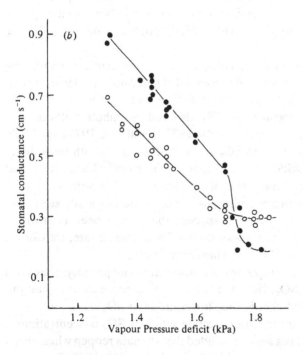

Fig. 2. Variation of stomatal conductance with vapour pressure deficit for plants of *Phaseolus vulgaris* (*a*) and *Vicia faba* (*b*) exposed either to charcoal-filtered air (open circles) or to air containing 35 ppb SO_2 (filled circles). From Black & Unsworth (1980).

exposures of up to one day. But Beckerson & Hofstra reported only partial recovery in clean air, and Black & Black (1979a) and Black & Unsworth (1980) found that responses persisted for at least several photoperiods. It seems most likely that the reversibility depends critically on the uptake of SO_2 and on the utilization and translocation of resulting products, factors which will vary with species, environment and stage of growth.

How does humidity affect responses to SO_2? Fig. 2 shows that there are substantial differences between the responses of species whose stomata are sensitive or insensitive to changes in vpd. In *Phaseolus vulgaris* L. (Fig. 2a), exposure to SO_2 resulted in an increase of conductance at all humidities; and the conductances of plants in either clean air or polluted air were not influenced by changing humidity. In contrast, stomatal conductance of *V. faba* (Fig. 2b) in clean air and in polluted air decreased with increasing vpd. At low vpd there was increased conductance when plants were exposed to SO_2; at high vpd, conductances were decreased by SO_2, confirming the responses observed at only two humidities by Majernik & Mansfield (1970). Similar patterns of SO_2 response have been found for other vpd sensitive species, *Helianthus annuus* L. and *Nicotiana tabacum* L, but with the vpd for 'changeover' from opening to closure differing between species (Black & Unsworth, 1980).

When plants are exposed to high SO_2 concentrations, particularly for long periods, stomatal closure is usually observed (Sij & Swanson, 1974; Bonte, De Cormis & Louguet, 1975; Menser & Heggestad, 1966; V. J. Black, unpublished). Since exposure to SO_2 also reduces photosynthesis and increases respiration (Black & Unsworth, 1979b; Hällgren, 1979), it is likely that stomatal responses to high SO_2 concentrations arise both from direct effects of sulphur dioxide on the turgor of the stomatal apparatus and indirectly *via* increased concentration of carbon dioxide within the leaf. Possibly because these two mechanisms of action are not easily separated and differ in importance between species, there have been contrasting interpretations of the mechanisms of stomatal response (Bonte, De Cormis & Louguet, 1975, 1977; Black & Unsworth, 1980).

Fig. 3 illustrates typical responses of transpiration and photosynthesis to a high concentration of SO_2. Note the brief initial increase in transpiration, consistent with stomatal opening for low uptakes of SO_2.

It is not clear whether the closure response to high SO_2 concentrations is reversible. Several authors have concluded that stomata reopen when plants are exposed to clean air (Bonte *et al.* 1975; Sij & Swanson, 1974; Taniyama, 1972), but it is also likely that non-stomatal water loss increases with time as a result of cellular and cuticular injury, and this could be mistaken for a recovery in stomatal transpiration.

Can stomatal closure reduce SO_2-induced injury? At present there is no direct evidence that closure is associated with resistance to chronic or acute damage, but Table 1 shows recent work by Bressan, Wilson & Filner (1978) suggesting that this may be so. Four cultivars of Cucurbitaceae were exposed to very high concentrations of SO_2 so that the dose (the product of concentration and time) was approximately the same for each cultivar. Leaf necrosis induced by SO_2 differed considerably between cultivars and correlated well with measured SO_2 uptake. Stomatal density and stomatal dimensions in clean air did not differ significantly between cultivars, and so it seems likely that the stomata of resistant cultivars closed sufficiently rapidly in response to SO_2 to limit the uptake and restrict the damage. It would be particularly interesting to know if similar responses occurred at low SO_2 concentrations, more typical of field exposure.

Sites of action of sulphur dioxide. If stomatal opening is the basic response to low concentrations of sulphur dioxide it could arise either from an increase in guard cell turgor or from a loss of turgor in adjacent epidermal cells. Recent evidence from Black & Black (1979a, b) indicates that, for *V. faba*, the latter mechanism is responsible. Table 2 indicates that epidermal strips taken from leaves of plants exposed to SO_2 and stained in neutral red

Fig. 3. Transpiration (filled circles) and net photosynthetic rates (open circles) of *Phaseolus vulgaris* exposed to 3000 ppb SO_2 for 1 h. Rates are expressed relative to those immediately prior to SO_2 treatment: $100 = 170 \pm 20 \, \text{g} \, H_2O \, \text{m}^{-2} \, \text{h}^{-1}$ and $1.7 \pm 0.2 \, \text{g} \, CO_2 \, \text{m}^{-2} \, \text{h}^{-1}$. Courtesy of Sij & Swanson (1974).

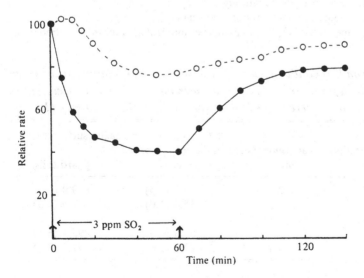

consistently showed lower viability of epidermal cells compared with controls from plants in clean air. In contrast, at relatively low SO_2 concentrations, survival of guard cells did not differ significantly between treatment and control. When the SO_2 exposure exceeded 175 ppb for more than 2 h, the viability of guard cells decreased; initially swelling of chloroplasts was observed (as reported by Wellburn, Majernik & Wellburn, 1972) and ultimately there was extensive disorganization of cell structure.

Light microscopy of epidermal strips cannot reveal whether epidermal cells are killed *in situ* by SO_2, or whether perhaps a weakening of cell membranes leads to structural damage only on taking the epidermal strip. However, scanning electron micrographs of intact leaf samples (Fig. 4) show clearly that epidermal cell turgor *is* lost *in situ* on exposure to SO_2.

Table 1. *$SO_{1 \cdot 2}$ dose, uptake and leaf necrosis in four cultivars of Cucurbitaceae. Calculated from results of Bressan, Wilson & Filner (1978)*

Cultivar[a]	Dose[b] gSO_2m^{-3} s	Uptake[c] $mgSO_2$ m^{-2}	Necrosis % leaf area	Necrosis Dose $\times 10^3$	Necrosis Uptake $\times 10^3$
1	412	219	52.8	128	241
2	480	131	32.9	69	251
3	480	104	23.1	48	222
4	532	90	17.6	33	196

[a]Cultivars: 1. *C. sativus* L. inbred line SC25
　　　　　　2. *C. sativus* L. cv. National Pickling
　　　　　　3. *C. pepo* L. cv. Prolific Straightneck Squash
　　　　　　4. *C. pepo* L. cv. Small Sugar Pumpkin
[b] Dose is the product of concentration and time.
[c] Uptake is the measured SO_2 flux to plants per unit leaf area for the duration of the experiment.

Table 2. *Mean leaf conductances and cell survival in epidermal strips taken from* Vicia faba *plants exposed either to clean air or to SO_2 at concentrations shown for 2 h. Strips were stained in neutral red. After Black & Black (1979a)*

SO_2 concentration ppb	Leaf conductance mm s^{-1}	% Cell survival on adaxial surface ($\pm$ s.e.) adjacent epidermal cells	guard cells
0	3.4	64.2 (2.5)	100.0 (0)
18	4.1	35.4 (1.0)	97.5 (2.5)
0	3.5	54.8 (2.5)	93.1 (3.1)
70	4.1	27.7 (2.8)	91.3 (3.0)

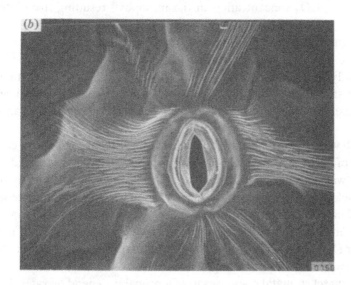

Fig. 4. Scanning electron micrographs of leaves of *Vicia faba* exposed either to 175 ppb SO_2 for 2 h (*a*) or to clean air (*b*). Note the turgid epidermal and guard cells in the clean air treatment, and the collapsed epidermal cells in the polluted treatments. Courtesy of C. R. Black & V. J. Black, unpublished photographs.

Our current hypotheses to explain stomatal responses to SO_2 in *V. faba* are as follows.

(i) At very low SO_2 concentrations (very slow rates of uptake) there may be weakening of membranes and subsequent loss of turgor in small numbers of epidermal cells. Stomatal opening results, but the response is reversible.

(ii) When SO_2 concentrations exceed about 20 ppb in our exposure cabinets, the increased rates of uptake of SO_2 further weaken membranes and lead to a loss of viability of an increasingly large proportion of epidermal cells. As a consequence, there is irreversible stomatal opening.

(iii) At SO_2 concentrations exceeding about 175 ppb, uptake of SO_2 causes swelling of guard cell chloroplasts, attacks guard cell membranes, and ultimately kills guard cells. A fall in guard cell turgor causes stomatal closure, irrespective of the turgor of epidermal cells (Glinka, 1971; Meidner & Bannister, 1979). Stomatal closure may also result from increases in intracellular CO_2 concentration in the mesophyll resulting from interference of SO_2 with photosynthetic pathways.

Epidermal cells may be attacked before guard cells by virtue of their greater exposure to SO_2 (Black & Black, 1979a). Since guard cells are usually protected by the cuticle, SO_2 probably has to reach them by entering stomata and then by transfer through adjacent epidermal cells, a route taken by other substances entering guard cells (Squire & Mansfield, 1972).

At cell walls within substomatal cavities of plants exposed to low concentrations of SO_2, the SO_2 concentration appears to be close to zero, indicating that the walls of mesophyll and epidermal cells are very good sinks for the gas (Black & Unsworth, 1979c). In such circumstances, SO_2 uptake would tend to follow the shortest diffusion path which is to adjacent epidermal cells and this may be another reason why stomata respond so rapidly to SO_2. Meidner (1976) first drew attention to this short diffusion path in relation to water loss from the stomatal apparatus.

Thus, direct stomatal responses to SO_2 probably depend on rates of entry of the gas into the leaf, sites of chemical reactions with tissue, and transfer pathways of toxic products. Further understanding is likely to come by studying how stomatal responses of whole plants depend on uptake, as determined by gas concentration and time. Studies at the cellular level, perhaps using labelled sulphur compounds, are also needed to assist in locating points of injury.

Oxides of nitrogen

The main sources of oxides of nitrogen (NO) are the combustion at high temperature of coal, oil and petroleum products. The most abundant oxides produced are NO and NO_2, but in the free atmosphere NO is usually rapidly oxidized to NO_2. In photochemical smogs in Los Angeles, arising from vehicle exhaust emissions, daily mean NO_x concentrations are often in the range 20–1000 ppb, with shorter peaks up to 4000 ppb (Stern, 1977). Capron & Mansfield (1975) pointed out that NO_x may be produced during CO_2 enrichment of glasshouses, and reported concentrations of up to 400 ppb, consisting predominantly (60–75%) of NO.

It is well established that NO_2 reduces photosynthesis at concentrations above about 500 ppb (Hill & Bennett, 1970; Tingey, Reinert, Dunning & Heck, 1971; Bull & Mansfield, 1974; Srivastava, Jolliffe & Runeckles, 1975 *a, b*); there is less information about responses to NO, but Hill & Bennett (1970) and Capron & Mansfield (1976) reported significant reductions in photosynthesis at concentrations above 600 ppb and 250 ppb, respectively.

Stomatal closure did not account for the reductions in photosynthesis of *Phaseolus vulgaris* observed by Srivastava *et al.*, who analysed detailed gas exchange measurements and showed that (i) stomatal conductance did not decrease until NO_2 concentrations exceeded 7000 ppb, (ii) there appeared to be a significant internal resistance to NO_2 uptake in the light, (iii) NO_2 increased the internal (mesophyll) resistance to CO_2 uptake. They concluded that stomatal closure at high concentrations of NO_2 was probably a consequence of elevated CO_2 concentration within the leaf. At high external CO_2 concentrations a decrease in photosynthesis and stomatal closure were still observed, and there is clearly a need for further study in such conditions with NO_2 and NO to determine responses that may occur in CO_2-enriched glasshouses.

Responses to mixtures of pollutants

When responses, sites of attack and mechanisms of action of *single* pollutants on vegetation are only poorly understood, it must seem rash to venture to review the tangled literature of responses to mixtures. However, in the natural environment pollutants seldom occur singly, and the identification of responses to mixtures is an important practical problem. So, undeterred, we present here two examples of such responses.

Ozone and sulphur dioxide

Fig. 5 shows measurements of Beckerson & Hofstra (1979*a*) of stomatal conductance in *Phaseolus vulgaris* exposed to SO_2 and O_2 at concentrations of 150 ppb, singly and as a mixture. Conductance increased in response to SO_2 alone, decreased to O_3 alone and decreased further in response to the mixture, especially after four days. Similar responses were also found for radish, cucumber and soybean (Beckerson & Hofstra, 1979b).

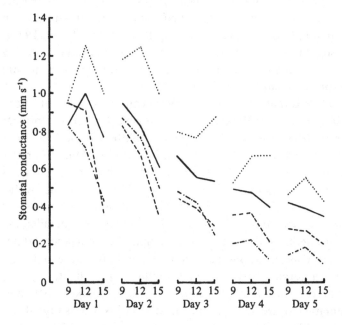

Fig. 5. Variation with time of the stomatal conductance of adaxial surfaces of primary leaves of *Phaseolus vulgaris* exposed in clean air (full line), 150 ppb SO_2 (dotted line), 150 ppb O_3 (interrupted line), and 150 ppb SO_2 + 150 ppb O_3 (dots and dashes). Recalculated from Beckerson & Hofstra (1979a).

The responses to the pollutants singly are consistent with those reviewed earlier but the mechanism is not known. It may be that brief stomatal opening by SO_2 results in an increased flux of ozone, leading to closure which is greater than in ozone alone. Or the combined action of SO_2 and O_3 on membranes may substantially reduce guard cell viability, leading to closure such as is observed at high SO_2 concentrations.

Finally, Fig. 6 shows fitted curves from one of the first field studies of stomatal responses to mixtures of pollutants (Coyne & Bingham, 1978). Plots of *P. vulgaris* were exposed to H_2S and O_3, singly and as mixtures at high and low concentrations of H_2S. The figure shows that with H_2S alone there was stomatal opening at 740 ppb but closure at 5030 ppb, responses similar to those for SO_2, but at much higher concentrations, possibly because of the relatively low solubility of H_2S. Ozone alone (72 ppb) induced stomatal closure, but there was greater closure in response to mixtures. Interpretation of these complex responses to mixtures must await a clearer understanding of responses to single gases.

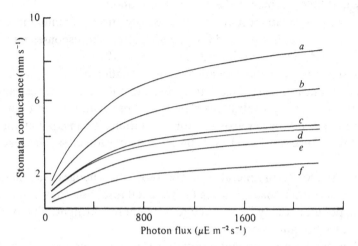

Fig. 6. The relation between leaf diffusive conductance and photon flux density for *Phaseolus vulgaris* plants growing in the field. Treatments: *a*, 740 ppb H_2S; *b*, clean air; *c*, 740 ppb H_2S + 72 ppb O_3; *d*, 72 ppb O_3; *e*, 5030 ppb H_2S; *f*, 5030 ppb H_2S + 72 ppb O_3. After Coyne & Bingham (1978).

Present and future

It is common for reviews to reach well-defined 'Conclusions', but our subject is not at such a stage. A few years ago it would have been difficult to decide, by reviewing the literature, whether there were any consistent stomatal responses to major pollutants; the situation now seems more clear-cut. Stomata generally open in response to low concentrations of SO_2 but close at high concentrations; H_2S elicits similar movements. Ozone usually induces stomatal closure, and responses to NO_2 seem to be small and associated with metabolic responses. Much less is known about sites of action of pollutants in the intact plant, although there has been considerable study of cellular and biochemical effects *in vitro*. There is a particular need for studies of effects on cells of the stomatal complex. Some of the questions still to be answered are given below.

How do responses depend on absorbed fluxes of pollutants? Can such knowledge help to explain reversible and irreversible responses? Are there species or varieties in which stomata close sufficiently rapidly and completely to restrict pollutant uptake and thus give protection from further damage? Can such characteristics be transferred by plant breeders?

Where are the sites of action of pollutants which induce stomatal responses, and what are the mechanisms of such action? Do sites and mechanisms differ between species? How does the environment for growth (laboratory and field) influence the action of pollutants?

How are passive stomatal responses influenced by effects of pollutants on photosynthesis and respiration, and how do such stomatal responses influence growth and development in field and laboratory?

Increased sensitivity of measurement and sophistication of techniques has enabled the identification, in the laboratory, of stomatal responses to certain pollutants at pollutant concentrations similar to those found out-of-doors. However, with natural species, exposed for generations to some pollutants, evolution of resistance is likely. How do the stomata of evolved species react to pollutants?

Stomatal physiologists in general should be aware that changes in ambient pollutant concentrations may influence the stomatal responses they observe – control of pollutants may be as necessary as control of light, CO_2 and humidity in experimental work!

Stomatal responses to pollutants represent a new and underdeveloped area of stomatal physiology. Techniques developed in other areas reviewed in this book will undoubtedly play a part in future developments of this particular topic. Conversely, the ability of pollutants to modify stomatal behaviour may give the physiologist exciting new possibilities for manipulating stomata in a wide range of experimental work.

References

Adedipe, N. O., Khatamian, H. & Ormrod, D. P. (1973). Stomatal regulation of ozone phytotoxicity in tomato. *Zeitschrift für Pflanzenphysiologie* **68**, 323–8.

Ashenden, T. W. (1979). Effects of SO_2 and NO_2 pollution on transpiration in *Phaseolus vulgaris* L. *Environmental Pollution* **18**, 45–50.

Ashenden, T. W. & Mansfield, T. A. (1977). Influence of wind speed on the sensitivity of ryegrass to SO_2. *Journal of Experimental Botany* **28**, 729–35.

Beckerson, D. W. & Hofstra, G. (1979a). Stomatal responses of white bean to O_3 and SO_2 singly or in combination. *Atmospheric Environment* **13**, 533–5.

Beckerson, D. W. & Hofstra, G. (1979b). Response of leaf diffusive resistance of radish, cucumber and soybean to O_3 and SO_2 singly or in combination. *Atmospheric Environment* **13**, 1263–8.

Biscoe, P. V., Unsworth, M. H. & Pinckney, H. R. (1973). The effects of low concentrations of sulphur dioxide on stomatal behaviour in *Vicia faba*. *New Phytologist* **72**, 1299–1306.

Black, C. R. & Black, V. J. (1979*a*). The effects of low concentrations of sulphur dioxide on stomatal conductance and epidermal cell survival in field bean (*Vicia faba* L). *Journal of Experimental Botany* **30**, 291–8.

Black, C. R. & Black, V. J. (1979*b*). Light and scanning electron microscopy of SO₂ induced injury to leaf surfaces of field bean *(Vicia faba L)*. *Plant, Cell, and Environment* **2**, 329–33.

Black, V. J. & Unsworth, M. H. (1979*a*). A system for measuring effects of sulphur dioxide on the gas exchange of plants. *Journal of Experimental Botany* **30**, 81–8.

Black, V. J. & Unsworth, M. H. (1979*b*). Effects of low concentrations of sulphur dioxide on net photosynthesis and dark respiration of *Vicia faba* L. *Journal of Experimental Botany* **30**, 473–83.

Black, V. J & Unsworth, M. H. (1979*c*). Resistance analysis of sulphur dioxide fluxes to *Vicia faba* L. *Nature* **282**, 68–9.

Black, V. J. & Unsworth, M. H. (1980). Stomatal responses to sulphur dioxide and vapour pressure deficit. *Journal of Experimental Botany* **31**, 667–77.

Bonte, J., De Cormis, L. & Louguet, P. (1975). Influence d'une pollution par le dioxyde de soufre sur le degre d'overture des stomates du *Pelargonium X hortorum*. *Comptes Rendus de L'Académie des Sciences, Paris D***280**, 2377–80.

Bonte, J., De Cormis, L. & Louguet, P. (1977). Inhibition en anaerobiose, de la reaction de fermeture des stomates du *Pelargonium* en presence de dioxyde de soufre. *Environmental Pollution* **12**, 125–33.

Brennan, E. & Leone, I. A. (1968). The response of plants to sulphur dioxide or ozone polluted air supplied at various flow rates. *Phytopathology* **58**, 1661–4.

Bressan, R. A., Wilson, L. G. & Filner, P. (1978). Mechanisms of resistance to sulphur dioxide in the Cucurbitaceae. *Plant Physiology* **61**, 761–7.

Bull, J. N. & Mansfield, T. A. (1974). Photosynthesis in leaves exposed to SO₂ and NO₂. *Nature* **250**, 443–4.

Capron, T. M. & Mansfield, T. A. (1975). Generation of nitrogen oxide pollutants during CO₂ enrichment of glasshouse atmospheres. *Journal of Horticultural Science* **50**, 233–8.

Capron, T. M. & Mansfield, T. A. (1976). Inhibition of net photosynthesis in tomato in air polluted with NO and NO₂. *Journal of Experimental Botany* **27**, 1181–6.

Coyne, P. I. & Bingham, G. E. (1978). Photosynthesis and stomatal light responses in snap beans exposed to hydrogen sulfide and ozone. *Journal of The Air Pollution Control Association* **28**, 1119–23.

Dugger, W. & Ting, I. P. (1970). Air pollution oxidants – their effects on metabolic processes in plants. *Annual Review of Plant Physiology* **21**, 215–34.

Evans, L. S. & Ting, I. P. (1974). Ozone sensitivity of leaves: relationship to leaf water content, gas transfer, gas transfer resistance and anatomical characteristics. *American Journal of Botany* **61**, 592–7.

Glinka, Z. (1971). The effect of epidermal cell water potential on stomatal response to illumination of leaf discs of *Vicia faba*. *Physiologia Plantarum* **24**, 476–9.

Hällgren, J.-E. (1978). Physiological and biochemical effects of sulphur dioxide on plants. In *Sulphur in the Environment: Part 2. Ecological*

Impacts, ed. J. O. Nriagu, pp. 163–209. New York: John Wiley and Sons.

Harrison, R. M. & Holman, C. D. (1979). The contribution of middle- and long-range transport of tropospheric photochemical ozone to pollution at a rural site in North-West England. *Atmospheric Environment* **13**, 1535–45.

Heath, R. L. (1975). Ozone. In *Responses of Plants to Air Pollution,* ed. J. B. Mudd & T. T. Kozlowski, pp 23–55. New York: Academic Press.

Hill, A. C. & Bennett, J. H. (1970). Inhibition of apparent photosynthesis by nitrogen oxides. *Atmospheric Environment* **4**, 341–8.

Hill, A. C. & Littlefield, N. (1969). Ozone. Effect on apparent photosynthesis, rate of transpiration and stomatal closure in plants. *Environmental Science and Technology* **3**, 52–6.

Keller, H. & Muller, J. (1958). Beitrage zur Erfassung der durch schweflige Saure hervorgerufenenRauchladen an Nadelholzern. *Forstwissenschaftliche Forschungen* **10**, 5–63.

Knudson Butler, L. & Tibbitts, T. W. (1979). Stomatal mechanisms determining resistance to ozone in *Phaseolus vulgaris* L. *Journal of the American Society of Horticultural Science* **104**, 213–16.

Majernik, O. & Mansfield, T. A. (1970). Direct effect of SO$_2$ pollution on the degree of opening of stomata. *Nature* **227**, 377–8.

Majernik, O. & Mansfield, T. A. (1971). Effects of SO$_2$ pollution on stomatal movements in *Vicia faba. Phytopathologische Zeitschrift* **71**, 123–8.

Mansfield, T. A. (1973). The role of stomata in determining the responses of plants to air pollutants. *Commentaries in Plant Science* **2**, 11–20.

Mansfield, T. A. & Majernik, O. (1970). Can stomata play a part in protecting plants against air pollutants? *Environmental Pollution* **1**, 149–54.

Martin, A. & Barber, F. (1973). Further measurements around modern power stations. 1. Observed ground level concentrations of sulphur dioxide. *Atmospheric Environment* **7**, 17–37.

Meidner, H. (1976). Water vapour loss from a physical model of a substomatal cavity. *Journal of Experimental Botany* **27**, 691–4.

Meidner, H. & Bannister, P. (1979). Pressure and solute potentials in stomatal cells of *Tradescantia virginiana. Journal of Experimental Botany* **30**, 255–65.

Menser, H. A. & Heggestad, H. E. (1966). Ozone and SO$_2$ synergism: injury to tobacco plants. *Science* **153**, 424–5.

Mudd, J. B. & Kozlowski, T. T. (ed.) (1975). *Responses of Plants to Air Pollution.* New York: Academic Press.

Mukammal, E. I. (1965). Ozone as a cause of tobacco injury. *Agricultural Meteorology* **2**, 145–65.

Rich, S. & Turner, N. C. (1972). Importance of moisture on stomatal behaviour of plants subjected to ozone. *Journal of the Air Pollution Control Association* **22**, 718–21.

Rich, S., Waggoner, P. E. & Tomlinson, H. (1970). Ozone uptake by bean leaves. *Science* **169**, 79–80.

Šesták, Z., Čatský, J. & Jarvis, P. G. (ed.) (1971). *Plant Photosynthetic Production: Manual of Methods.* The Hague: W. Junk.

Sij, J. W. & Swanson, C. A. (1974). Short-term kinetics on the inhibition of photosynthesis by sulphur dioxide. *Journal of Environmental Quality* **3**, 103–7.

Spedding, D. J. (1969). Uptake of sulphur dioxide by barley leaves at low sulphur dioxide concentrations. *Nature* **224**, 1229–30.

Squire, G. R. & Mansfield, T. A. (1972). A simple method of isolating stomata on detached epidermis by low pH treatment: Observations of the importance of subsidiary cells. *New Phytologist* **71**, 1033–43.

Srivastava, H. S., Jolliffe, P. A. & Runeckles, V. C. (1975*a*). Inhibition of gas exchange in bean leaves by NO₂. *Canadian Journal of Botany* **53**, 466–74.

Srivastava, H. S., Jolliffe, P. A. & Runeckles, V. C. (1975*b*). The effects of environmental conditions on the inhibition of leaf gas exchange by NO₂. *Canadian Journal of Botany* **53**, 475–82.

Stern, A. C. (ed.) (1977). *Air Pollution. Vol. 2. The effects of air pollution*, 3rd edn. New York: Academic Press.

Taniyama, T. (1972). Studies on the development of symptoms and the mechanism of injury caused by sulphur dioxide in crop plants. *Bulletin of the Faculty of Agriculture,* Mie University, Tsu, Japan, No. 44 11–130. (In Japanese with English summary.)

Tingey, D. T., Reinert, R. A., Dunning, J. A. & Heck, W. W. (1971). Vegetation injury from the interactions of NO₂ and SO₂. *Phytopathology* **61**, 1506–11.

Unsworth, M. H. (1981). The exchange of carbon dioxide and air pollutants between vegetation and the atmosphere. In *Plants and their Atmospheric Environment,* ed. J. Grace, D. Ford & P. G. Jarvis, pp. 111–38. Oxford: Blackwell.

Unsworth, M. H., Biscoe, P. V. & Pinckney, H. R. (1972). Stomatal responses to sulphur dioxide. *Nature* **239**, 458–9.

Verkroost, M. (1974). The effect of ozone on photosynthesis and respiration of *Scenedesmus obtrusiusculus* Chod. with a general discussion of effects of air pollutants in plants. *Medelingen van de Landbouwhogeschool te Wageningen* **74**, 1–78.

Vukovich, F. M., Bach, W. D., Crissman, B. W. & King, W. J. (1977). On the relationship between high ozone in the rural surface layer and high pressure systems. *Atmospheric Environment* **11**, 967–83.

Wellburn, A. R., Majernik, O. & Wellburn, F. A. M. (1972). Effects of SO₂ and NO₂ polluted air on the ultrastructure of chloroplasts. *Environmental Pollution* **3**, 37–49.

P. G. AYRES

Responses of stomata to pathogenic microorganisms

Introduction

Each plant has its own characteristic group of pathogenically-induced diseases to which it may succumb, each disease having a different aetiology. Luckily for us, the physiological responses of plants to infection are limited in number. Where stomata are concerned it seems reasonable to make the generalization that the predominant response to infection is to lose the ability to respond to variation in those environmental factors to which, as demonstrated in earlier chapters, they are normally so sensitive. Thus, the amplitude of stomatal movements declines with time after infection, although the mean position about which oscillations occur may represent either a progressively more open or more closed state.

Altered stomatal behaviour is only an inconsequential part of the syndrome in those diseases where the pathogen causes rapid and widespread death of tissues, e.g. 'Damping-off' of seedlings caused by *Pythium* spp. However, in the majority of diseases, where the progress of infection is slower, altered stomatal behaviour can affect fluxes of carbon dioxide and water to and from living tissue and, hence, the continued growth and survival of the plant. In these cases the impact of altered stomatal behaviour is dependent upon the speed with which alterations occur, and the extent to which the influence of the pathogen reaches beyond the tissue it has colonized, as well as the nature of the changes. Unfortunately, information on these points is usually incomplete. Inferences may be drawn, and our knowledge of the mechanisms by which pathogens bring about changes often helps in this direction, but for many diseases the importance of altered stomatal behaviour has not been clearly demonstrated.

Stomata in pathological processes

Before considering the responses of stomata to pathogens it is useful to look briefly at the physical relationship between stomata and the development of pathogens. In root diseases and many shoot diseases the pathogen never directly challenges tissue bearing stomata, but even when the main challenge is to the aerial parts of the plant the stomatal apparatus is remarkably resistant to direct attack. Guard cells are invaded directly by only a few pathogens, e.g. by *Lophodermium pinastri* (Schrad.) Chev. (Jones, 1935) which causes leaf-cast of pine, while subsidiary cells arc also generally more resistant to attack than are other epidermal cells, though Hirata (1971) found that under certain conditions 80–90% of initial penetrations of barley leaves by *Erysiphe graminis* D.C. f.sp. *hordei* (powdery mildew) involved subsidiary cells which together occupied only 3% of the total leaf surface. The stomatal pore provides an opening which is the natural entrance to the plant for a wide variety of bacteria and fungi. Bacteria, which are incapable of forced, active invasion, pass through the pore in a film of water before proliferating in the substomatal chamber; enzymes and toxins may be produced to disrupt the structure and metabolism of surrounding host cells. The requirement for water in the dispersion and invasion phases means that bacterial leaf diseases, such as Fire Blight of apples and pears

Fig. 1. Urediospore of rust, *Puccinia hordei*, infects a barley leaf. (*a*) The germ tube forms an appressorium, a, over guard cells, g. Infection thread passes between guard cells and differentiates a substomatal vesicle, sv. (*b*) Appressorium collapsed, branching mycelium penetrates mesophyll cells and forms haustoria, h. After Wheeler (1969).

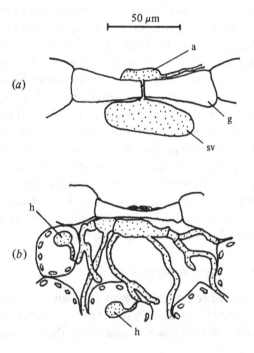

(*Erwinia amylovora* (Burrill) Winslow) and Wildfire of tobacco (*Pseudomonas tabacci* (Wolf & Foster) Stapp), are commonly associated with periods of high rainfall or atmospheric humidity.

Pathogenic fungi often take the opportunity the stomatal pore allows of reaching the substomatal chamber, an environment buffered from the extremes of insolation and desiccation found at the leaf surface, without having to invade host cells and thereby encounter their resistance mechanisms. Some fungi have an absolute requirement for this mode of entry. The fungus may actively seek the pore through swimming movements of motile zoospores, e.g. *Pseudoperonspora humili* (Miyabe & Tak.) G. W. Wilson and *Plasmopara viticola* Berl. & de Toni, downy mildews of hop and grape respectively, or through directed growth of the spore's germination tube which may be orientated towards the pore by chemical or topographic factors at the stomatal site, e.g. urediospores of many rust fungi (Family Uredinales) (Fig. 1).

It has sometimes been argued (see Royle, 1976 for a full discussion) that the size of the stomatal apparatus and its state of opening can restrict the passage of a penetrating fungal hypha, so acting as a resistance mechanism.

Fig. 2. Sporangiophores, s, of blight, *Phytophthora infestans*, emerge through stomatal pore of potato leaf and release zoosporangia z. After Webster (1970).

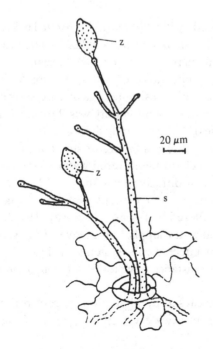

z

20 μm

z

s

Hence *Puccinia graminis tritici* Erikss. & Henn., wheat stem rust, normally only enters stomatal pores in the light, because it is inhibited by high carbon dioxide levels (Yirgou & Caldwell, 1963), but can be prevented from doing so by stomatal closure caused by water stress (Burrage, 1970).

Finally, it is noteworthy that the stomatal pore may serve as a means of exit from the leaf for the pathogen. Thus, sporangiophores of *Phytophthora infestans* (Mont.) de Bary (potato blight) extend through and beyond the stomatal pores before sporulation (Fig. 2).

The causes of abnormal stomatal behaviour

Water supply to the leaf is interrupted

Perhaps the simplest systems are those in which a soil-inhabiting pathogen attacks the roots or stem-base of a plant, reducing the supply of water to the upper parts of the plant and thereby inhibiting full stomatal opening. The supply of water may be reduced because the pathogen kills host cells, so reducing the absorptive area of the root, or causes solute leakage from host cells, so destroying osmotic gradients along which water is taken up, or blocks or embolizes conducting elements, or produces a combination of these effects. Thus, wilting is a characteristic of many root infections even though the disease may be most prevalent under humid or wet conditions.

Both root-rot of safflower, caused by *Phytophthora drechsleri* Tucker, and black-root of sugar beet, caused by *Aphanomyces cochlioides* Drechsl., are transmitted by zoospores which need free water in which to swim, yet both diseases are characterized by the appearance of wilt symptoms 5–7 days after infection. In sugar beet the diffusive resistance of a disease-susceptible variety was $4.5\,s\,cm^{-1}$, and that of a resistant variety was $2.6\,s\,cm^{-1}$, compared with a healthy control value of $1.6\,s\,cm^{-1}$ only five days after infection (Safir & Schneider, 1976). In both diseases failure of stomata to open in infected plants was attributable solely to the lowered water content of the leaf since the relationships between diffusive resistance and leaf relative water content, and between relative water content, water potential and solute potential (Fig. 3) were unaltered by disease (Duniway, 1975). Two factors contributed to the reduced water supply to the leaves. Firstly, infection reduced the fresh weight of infected root systems by 43–56%, and, secondly, infection increased the resistance to water flow through the roots and stem.

Effects on stomatal behaviour caused by Phytophthora root rot of safflower are similar to those seen in a wide variety of plants infected by vascular wilt fungi. The latter fungi, although confined to the vascular system of the plant, spread upward through the stem and even into the leaves of the plant. Extensive investigation of Verticillium wilt (*V. dahliae* Kleb) of Chrysanthemum by Canadian workers indicates that reduced stomatal opening is evident eight days before symptoms appear (MacHardy, Busch & Hall, 1976). The leaf diffusion resistance becomes progressively higher (MacHardy, Hall & Busch, 1974) and as symptoms become apparent there is a strong correlation between diffusive resistance, relative water content and the degree of colonization of different lobes of the leaf by the fungus (Fig. 4). Interruption of the flow of water-soluble dyes and radioactive tracers through diseased leaves shows readily how the fungus interferes with the supply of water to distal regions of the plant.

Changes in the levels of hormones that exert regulatory effects on stomatal behaviour, i.e. cytokinins and abscisic acid, may accompany

Fig. 3. Phytophthora rot of safflower does not alter the relationship between abaxial leaf diffusive resistance and water potential (interrupted line; open circles, healthy, filled circles, diseased), or between leaf water and solute potentials (full line; individual points not shown). From Duniway (1975).

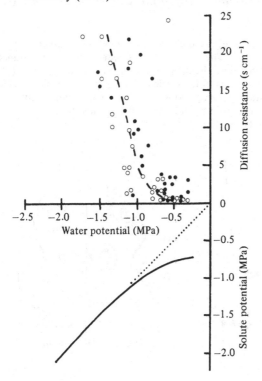

pathogen-induced wilt. Excepting those vascular wilts where chlorosis and other symptoms of senescence *precede* wilting, e.g. Verticillium wilt of sunflower (*V. dahliae*), it is generally agreed that altered hormone levels are a result rather than a primary cause of the syndrome. Where more destructive root pathogens are concerned it is possible that, if disease progress is unusually slow, selective destruction of hormone producing sites in the root could have an effect on stomatal behaviour independent of that caused by an interrupted flow of water.

Toxins are produced by the pathogen

The term toxin is applied to a wide variety of non-enzymic substances which are elaborated by microorganisms and interfere with the normal metabolism of the host. Their diversity, and our imperfect knowledge of their chemistry, makes classification by effect on the plant rather than by molecular structure the normal rule. In spite of numerous reports that particular toxins affect water relations of host (or, too often non-host) plants and, therefore, by implication stomatal behaviour, evidence is rare that effects of toxins on stomatal behaviour play a central role in the development of the disease syndrome.

A number of species of *Helminthosporium* produce toxins of differing

Fig. 4. Water stress develops in chrysanthemum leaves infected by *Verticillium dahliae*. (*a*). The relationship between diffusive resistance and symptom development, healthy lobe tip (open squares), base (filled squares); diseased lobe tip (filled circles), base (open circles). (*b*). The relative water content (above) and number of fungal colonies cultured from 1 cm diameter leaf discs (below). Left half infected, right half control. From MacHardy *et al* (1974, 1976).

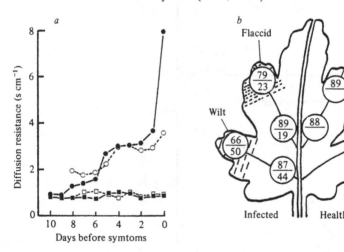

levels of host specificity and cause various leaf and ear blights of cereal crops in which changes in membrane permeability and altered water relations are noted. *H. maydis* Nisikado & Miyake Race T causes Southern Corn Leaf Blight of maize varieties carrying a particular cytoplasmically inherited mutation; both the pathogen and its toxin(s) (Race T toxin, see Table 1) are host specific. When supplied to excised maize leaves of a susceptible strain, the toxin severely inhibits transpiration and photosynthesis within 60 min of application by inhibiting light-induced potassium ion uptake by guard cells (Arntzen, Haugh & Bobick, 1973). The toxin inhibits potassium-stimulated ATP-ase activity in susceptible tissues (Tipton, Mondal & Benson, 1975) and generally causes solute leakage (Gracen, Grogan & Forster, 1972). The behaviour of stomata in naturally infected leaves is unknown but concentrations of toxin in the locality of stomata may be high, for a high proportion of first penetrations take place between guard and subsidiary cells, possibly because of breaks in the cuticle in this region (Wheeler, 1977). In this and similar diseases where a toxin has multiple sites of action there remains considerable doubt as to whether inhibition of stomatal opening *per se*, or more simply the destruction of a large portion of the leaf area of the plant, causes the stunted growth that results from infection.

We can be more certain that fusicoccin, a toxin produced by the fungus *Fusiccocum amygdali* Del. (Table 1), has a significant effect on stomatal behaviour in the natural development of Canker disease of peach and almond. The toxin has been isolated from naturally infected hosts (Ballio *et al.*, 1976) and upon reapplication to the natural host stimulates stomatal opening in the light and, more particularly, in the dark (Turner & Graniti, 1976) thus reproducing the stomatal opening in the dark that is a feature of

Table 1. *Chemical nature of some fungal toxins affecting stomatal behaviour*

Common name of toxin	Producer	Chemical nature
Race T toxin	*Helminthosporium maydis*	A mixture of three or possibly four tetracyclic triterpenoids, one in glycosidic form.
Fusicoccin	*Fusicoccum amygdali*	Carbotricyclic terpene in glucosidic form.
Tentoxin	*Alternaria tenuis*	Cyclic peptide (-L-leucyl-N-methyl-(Z)-dehydrophenyl-alanyl-glycyl-N-methyl-L-alanyl-).

leaves on infected branches (Table 2). The fungus enters the host through bud or leaf scars but spreads only slowly. However, wilt symptoms are seen at distances of up to 40 cm in advance of the fungus because the toxin moves ahead in the transpiration stream.

The toxin, unlike the fungus, is not host specific and this has meant that most of the work on the elucidation of the mechanism of action of fusicoccin has been carried out on species that are grown more conveniently than peach or almond. Both histochemical and X-ray microprobe analyses have shown that the toxin acts by stimulating potassium ion uptake by guard cells, though this is probably the result of a more fundamental property, that of stimulating proton flux and establishing gradients of potential energy across membranes (Marrè, 1979). It is interesting that tentoxin (Table 1), a similarly non-host specific toxin, produced by the fungus *Alternaria tenuis* Nees has the reverse property, that of inhibiting stomatal opening by inhibiting light-stimulated K^+ uptake by guard cells (Table 3) (Durbin, Uchytil & Sparapano, 1973). The activity of tentoxin seems to lie in its ability to inhibit photophosphorylation (Arntzen, 1972) and, thus, the supply of ATP for potassium accumulation. *A. tenuis* infects cotyledons of many dicotyledonous plants, as well as causing Brown Spot of leaves, but in both cases

Table 2. *Abaxial stomatal resistance of an almond leaf from a branch inoculated 16 days previously with* Fusicoccum amygdali *or sterile water (control). After Turner & Graniti (1976)*

| | Stomatal resistance s cm^{-1} | |
	F. amygdali	Control
Mid-day	14	11
15 min after sunset	16	67

Table 3. *Effect of 100 μM tentoxin on stomatal aperture of broad bean* Vicia faba. *After Durbin, Uchytil & Sparapano (1973)*

Treatment	Detached leaflet (aperture, μm)	Epidermal strip (aperture, μm)	Potassium accumulation by guard cell
Light	10.4 ± 2.9	11.5 ± 2.0	Accumulation
Light + Tentoxin	2.5 ± 3.2	2.8 ± 3.3	No accumulation
Dark	0.2 ± 0.7	_[a]	No accumulation
Dark + CO$_2$ free air	16.1 ± 2.8	_[a]	Accumulation
Dark + CO$_2$ free air + Tentoxin	4.9 ± 2.7	_[a]	Little accumulation

[a] no value given

chlorosis followed by necrosis is the characteristic symptom recorded and disturbance of stomatal behaviour has not been reported.

The pathogen invades the leaf

Systems in which it is most difficult to discern the causes of abnormal stomatal behaviour are those involving foliar pathogens. To some extent the pathogen disrupts the structure of the leaf, and competes with host cells for water and nutrients; it often liberates, or causes to be liberated from the host, substances that alter the permeability of host membranes to water and solutes. As part of these changes the pathogen can inhibit photosynthetic activity in the host so that the energy supply for ion transport and stomatal movement is diminished. The pathogen frequently destroys the integrity of the leaf cuticle, reducing resistance to non-stomatal water loss and, incidentally, preventing the investigator from inferring stomatal behaviour from measurements of water vapour diffusion rates.

Infection stimulates stomatal opening in the light in a few diseases. In potato blight, enhanced opening occurs in a band of tissue which surrounds the necrotic centre of lesions and which is in the process of being colonized by the fungus (Fig. 5) (Farrell, Preece & Wren, 1969). This illustrates how the effects of a pathogen are intimately related to its temporal development and, particularly in the case of a foliar pathogen, to its spatial development.

Fig. 5. Effect of development of blight lesions, *Phytophthora infestans*, on stomatal aperture in potato, and the modifying influence of leaf water deficit. Open circles, 0% deficit; filled triangles, 11% deficit; open triangles, 38% deficit. From Farrell *et al* (1969).

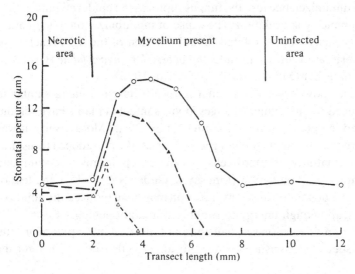

In areas of enhanced stomatal opening, the solute potential of guard cells approached $-5.0\,\text{MPa}$; this compares with values of $-1.9\,\text{MPa}$ and $-1.6\,\text{MPa}$ for guard cells from healthy tissue in light and dark respectively. No measurements were made on epidermal cells but it would seem that infection stimulates stomatal opening by altering the turgor relations between guard and epidermal cells.

Altered turgor relations between guard and epidermal cells are found in barley leaf blotch disease (*Rhynchosporium secalis* Oudem. J. J. Davis) where increased stomatal opening in the light occurs in the early stages of infection (Ayres, 1972). At this point in its development the fungus is confined to a subcuticular position in the leaf and causes permeability changes in surrounding host cells, possibly to supplement its nutrient supply (Jones & Ayres, 1972). Epidermal cells are affected more than guard cells and increased opening results for, as Glinka (1971) has pointed out, the same difference in turgor between guard and epidermal cells can lead to larger stomatal apertures at low than at high turgor pressures in epidermal cells. Thus, when infection has increased stomatal aperture from $6\,\mu\text{m}$ in healthy leaves to $15\,\mu\text{m}$ in infected leaves 18 days after inoculation, it has reduced epidermal cell turgor pressure by 36% but only reduced the difference between guard and epidermal cells by 15%.

Rust fungi, whose entry through stomatal pores has already been described, develop mainly in the intercellular spaces of the leaf and insert specialized absorptive organs, haustoria, into mesophyll cells (Fig. 1*b*). Stomatal movements are progressively inhibited but, in contrast to potato blight and barley leaf blotch diseases, the stomata become fixed in an almost closed position (Duniway & Durbin, 1971). Any reduction in water loss at this stage is minimized because the fungus ruptures the epidermis and cuticle before commencing sporulation. The cause of immobilization of stomata is unknown, but may well be related to an inhibition of the light reaction of photosynthesis, as occurs in rust infected beans (*Uromyces fabae* (Grev.) de Bary/*Vicia faba* L.) (Montalbini & Buchanan, 1974).

Viruses and powdery mildew fungi have effects on stomata similar to those produced by rust fungi. Changes in virus infected plants have seldom been studied in detail, but in maize infected by Dwarf Mosaic Virus it was noted that reduced stomatal opening was associated with reduced transpiration and the inability of guard cells of isolated epidermis to accumulate potassium ions on illumination (Lindsey & Gudauskas, 1975). The loss of photosynthetic pigments caused by infection may have reduced the ability of the leaf to trap enough energy to support stomatal opening.

In powdery mildew diseases, haustoria are inserted into epidermal cells, and under special circumstances into subsidiary cells (see p. 206) but the mycelium of the fungus remains on the surface of the leaf. Examination of the solute relations of the epidermis of powdery mildew infected leaves reveals that in pea, at least, although both guard and epidermal cells lose solutes, the main effect of disease is to prevent solute accumulation by guard cells upon illumination (Table 4). In pea, as in barley and oak mildews (Ayres, 1979), reductions in net photosynthesis occur sooner after infection than reductions in stomatal aperture (Ayres, 1976) (Fig. 6). As suggested for virus diseases, it may be that a loss of photosynthetic activity is the *cause* of stomatal malfunction because insufficient energy is supplied for active solute accumulation upon illumination. However, it is probably naïve to assume that stomatal behaviour is altered by a single factor. In addition, stomata might respond in part to enhanced carbon dioxide concentrations in their environment. These result not only from diminished photosynthetic uptake, but from increased rates of dark respiration and photorespiration and, also, respiratory activity of the pathogen.

Stomatal opening may be inhibited by substances synthesized by the host in response to infection. Thus, pisatin (a pterocarpan) which accumulates in mildewed peas is both fungi- and phytotoxic at high concentrations but at non-lethal levels, 1×10^{-4}M, may reduce stomatal opening (Fig. 7). Although pisatin is produced specifically by pea and its close relatives, other phytoalexins (post-infectionally produced resistance compounds) produced by different plants, but sharing with pisatin the ability to disrupt membrane permeability, may affect stomatal behaviour.

Table 4. *Effect of powdery mildew,* Erysiphe pisi, *on the solute relations of pea epidermis, and the ability of guard cells to accumulate solutes upon illumination*

| | | Solute potential[a], MPa | | |
		Guard cell	Epidermal cell	Difference
Healthy	Dark	− 1.40	− 0.93	0.47
	Light	− 1.96	− 0.96	1.00
	Change on illumination	0.56	0.03	0.53
Infected	Dark	− 1.06	− 0.67	0.39
	Light	− 1.19	− 0.79	0.40
	Change on illumination	0.13	0.12	0.01

[a] Solute potentials were measured by determining the point of 50% plasmolysis in mannitol. Unpublished data of Ayres.

Fig. 6. Effects of powdery mildew of pea, *Erysiphe pisi*, on gross photosynthesis, determined in 2% oxygen, open circles, healthy; filled circles, infected), and stomatal opening (open squares, healthy; filled squares, infected) in light and dark. From Ayres (1976).

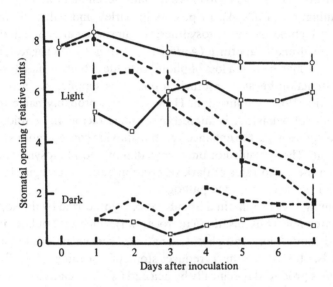

Fig. 7. Effects of 1×10^{-4}M pisatin on the diffusive resistance of pea leaves (open circles, control leaflet; filled circles, treated opposite leaflet). Differences after 2 h are significant at the 5% probability level. Ayres, unpublished data.

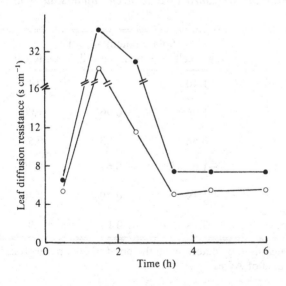

Foliar pathogens are usually confined to discrete lesions and their effects are localized (see Fig. 4) unless the host–parasite interaction produces a diffusible metabolite with the ability to regulate metabolism, for example the toxins discussed earlier (p. 211). However, recent observations show that powdery mildew lesions may be associated with increased stomatal opening in healthy areas of the same infected leaf (Ayres & Zadoks, 1979) (Fig. 8), or in other, healthy leaves of the infected plant (unpublished results of Williams in this laboratory). Whether this effect is mediated through the water and solute relations of the plant is not known yet but the important question is raised – to what extent can healthy areas of the plant compensate for loss of activity in diseased areas?

Consequences of altered stomatal behaviour

Water relations

The importance of altered stomatal behaviour depends very much upon the particular conditions under which each disease normally occurs. Thus, increased stomatal opening causes wilting and desiccation of leaves on peach and almond trees infected by *F. amygdali* which grow typically in a warm dry Mediterranean climate, but it rarely leads directly to such effects in blighted potatoes, or Rhynchosporium-infected barley, because both of

Fig. 8. Effect of powdery mildew, *Erysiphe graminis hordei*, on stomatal behaviour in flag leaves of 55-day-old barley plants (open squares, leaf of uninfected plant; half-filled squares, uninfected region of infected leaf; filled squares, infected region). From Ayres & Zadoks (1979).

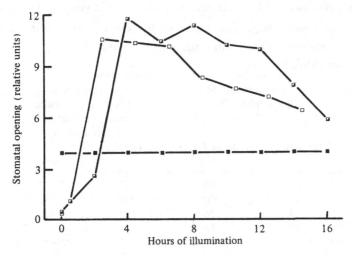

these diseases occur under cool wet conditions. However, in blighted potatoes a localized zone of increased photosynthesis occurs in the area occupied by abnormally open stomata (Farrell *et al.*, 1969), while in Rhynchosporium-infected barley transpiration and the accumulation of solute, ^{86}Rb, absorbed through the root increase in the affected area (Ayres & Jones, 1975). Both these changes may affect the nutrition of the host and parasite, and the balance of their relationship.

Decreased stomatal opening conserves water, but in root and vascular diseases this is often not enough to maintain the normal water balance of root and stem tissues, at least in susceptible plants, because the supply of water to the shoot is significantly diminished although soil water may be plentiful. In rust diseases any conservation of water that results from reduced stomatal opening lasts only until the epidermis is ruptured. Reduced stomatal opening does have important repercussions with respect to powdery mildew diseases, for amongst fungi these pathogens are remarkably tolerant of warm dry conditions. Reduced water loss from infected leaves makes more water available for uninfected leaves on the same plant which is beneficial, particularly in times of drought. On the other hand, higher leaf water potentials may favour the development of the pathogen (Ayres, 1977) and lead to more widespread infection. The impact of disease on plant water relations is obviously complex, and is discussed more fully elsewhere (Ayres, 1978).

Carbon dioxide exchange

Available evidence suggests that, by reducing the movement of carbon dioxide into leaves, reduced stomatal opening contributes to the reduction in rates of carbon dioxide fixation per unit leaf area which commonly result from disease. However, altered stomatal behaviour is probably

Table 5. *Tolerance to the Sugar Beet Yellows Virus among varieties of sugar beet and the maintenance of low leaf resistance to gas transport. After Hall, Hunt & Loomis (1972)*

	% Yield loss due to BYV infection		Resistance of leaf measured in a viscous flow porometer, relative units.	
Variety	Root Sugar	Root Fresh Wt	Healthy	Effects of infection
USH 20	52	43	0.79	+ 2.69
USH 7	46	40	1.40	+ 1.03
Y904	31	26	0.95	+ 0.06
813	20	16	0.91	+ 0.32

not the primary cause of the reduction in photosynthesis – indeed altered stomatal behaviour may be a result of changes taking place in chloroplasts. In rust and mildew diseases changes in photosynthetic efficiency appear *at least* as soon after infection as reduced stomatal opening (for full references see Ayres, 1979), while simultaneous increases in leaf resistance (r_1) and mesophyll (or internal) resistance (r_{mes}) to carbon dioxide transport are measured in Fusarium vascular wilt of tomato (Duniway & Slatyer, 1971), Peach Rosette and Decline Virus disease (Smith & Neales, 1977) and Sugar Beet Yellows Virus disease (Hall & Loomis, 1972). The relative importance of the two resistances varies according to growth conditions as well as the stage of infection in particular diseases. Hall, Hunt & Loomis (1972) have pointed out an interesting correlation between the continued growth of different sugar beet varieties infected by yellows virus and their maintenance of low leaf resistances to the transport of gases during infection (Table 5), but, generally, the importance of the consequences of altered stomatal behaviour in diseased plants should not be overstated.

References

Arntzen, C. J. (1972). Inhibition of photophosphorylation by tentoxin, a cyclic tetrapeptide. *Biochimica et Biophysica Acta* **283**, 539–42.

Arntzen, C. J., Haugh, M. F. & Bobick, S. (1973). Induction of stomatal closure by *Helminthosporium maydis* pathotoxin. *Plant Physiology* **52**, 569–74.

Ayres, P. G. (1972). Abnormal behaviour of stomata in barley leaves infected with *Rhynchosporium secalis* (Oudem.) J. J. Davis. *Journal of Experimental Botany* **23**, 683–91.

Ayres, P. G. (1976). Patterns of stomatal behaviour, transpiration, and CO_2 exchange in pea following infection by powdery mildew (*Erysiphe pisi*). *Journal of Experimental Botany* **27**, 354–63.

Ayres, P. G. (1977). Effects of leaf water potential on sporulation of *Erysiphe pisi* (pea mildew). *Transactions of the British Mycological Society* **68**, 97–100.

Ayres, P. G. (1978). Water relations of diseased plants. In *Water Deficits and Plant Growth* Vol. V. Ed. T. T. Kozlowski, pp. 1–60. London: Academic Press.

Ayres, P. G. (1979). CO_2 exchanges in plants infected by obligately biotrophic pathogens. In *Photosynthesis and Plant Development* ed. R. Marcelle, H. Clijsters & M. Van Poucke, pp. 343–54. London: Dr W. Junk.

Ayres, P. G. & Jones, P. (1975). Increased transpiration and the accumulation of root absorbed ^{86}Rb in barley leaves infected by *Rhynchosporium secalis* (leaf blotch). *Physiological Plant Pathology* **7**, 49–58.

Ayres, P. G. & Zadoks, J. C. (1979). Combined effects of powdery mildew disease and soil water level on the water relations and growth of barley. *Physiological Plant Pathology* **14**, 347–67.

Ballio, A., D'Alessio, V., Randazzo, G., Bottalico, A., Graniti, A.,
Sparapano, L., Bosnar, B., Casinovi, C. G. & Gribanovski-Sassu, O.
(1976). Occurrence of fusicoccin in plant tissues infected by
Fusicoccum amygdali Del. *Physiological Plant Pathology* 8, 163–9.

Burrage, S. W. (1970). Environmental factors influencing the infection
of wheat by *Puccinia graminis*. *Annals of Applied Biology* 66,
429–40.

Duniway, J. M. (1975). Water relations in safflower during wilting
induced by Phytophthora root rot. *Phytopathology* 65, 886–91.

Duniway, J. M. & Durbin, R. D. (1971). Some effects of *Uromyces
phaseoli* on the transpiration rate and stomatal response of bean
leaves. *Phytopathology* 61, 114–19.

Duniway, J. M. & Slatyer, R. O. (1971). Gas exchange studies on the
transpiration and photosynthesis of tomato leaves affected by
Fusarium oxysporum f.sp. *lycopersici*. *Phytopathology* 61, 1377–81.

Durbin, R. D., Uchytil, T. F. & Sparapano, L. (1973). The effect of
tentoxin on stomatal aperture and the potassium content of guard
cells. *Phytopathology* 63, 1077–8.

Farrell, G. M., Preece, T. F. & Wren, M. J. (1969). Effects of infection
by *Phytophthora infestans* (Mont.) de Bary on stomata of potato
leaves. *Annals of Applied Biology* 63, 265–75.

Glinka, Z. (1971). The effect of epidermal cell water potential on
stomatal response to illumination of leaf discs of *Vicia faba*.
Physiologia Plantarum 24, 476–9.

Gracen, V. E., Grogan, C. O. & Forster, M. J. (1972). Permeability
changes induced by *Helminthosporium maydis*, race T, toxin.
Canadian Journal of Botany 50, 2167–70.

Hall, A. E., Hunt, W. F. & Loomis, R. S. (1972). Variations in leaf
resistances, net photosynthesis and tolerance to the beet yellows virus
among varieties of sugar beet (*Beta vulgaris* L.) *Crop Science* 12,
558–60.

Hall, A. E. & Loomis, R. S. (1972). An explanation of the difference in
photosynthetic capabilities of healthy and beet yellows virus-infected
sugar beets (*Beta vulgaris* L.). *Plant Physiology* 50, 576–80.

Hirata, K. (1971). Fine structure and metabolic function of intracellular
fungal structures. In *Morphological and Biochemical Events in
Plant–Parasite Interaction*, ed. S. Akai & S. Ouchi, pp. 207–28.
Tokyo: The Phytopathological Society of Japan.

Jones, P. & Ayres, P. G. (1972). The nutrition of the subcuticular
mycelium of *Rhynchosporium secalis* (barley leaf blotch); permeability
changes induced in the host. *Physiological Plant Pathology* 2, 383–92.

Jones, S. G. (1935). The structure of *Lophodermium pinastri* (Schrad.)
Chev. *Annals of Botany* 49, 699–728.

Lindsey, D. W. & Gudauskas, R. T. (1975). Effects of maize dwarf
mosaic virus on water relations of corn. *Phytopathology* 65, 434–40.

MacHardy, W. E., Busch, L. V. & Hall, R. (1976). Verticillium wilt of
chrysanthemum; quantitative relationship between increased stomatal
resistance and local vascular disfunction preceding wilt. *Canadian
Journal of Botany* 54, 1023–34.

MacHardy, W. E., Hall, R. & Busch, L. V. (1974). Verticillium wilt of
chrysanthemum; relative water content and protein, R.N.A., and
chlorophyll levels in leaves in relation to visible wilt symptoms.
Canadian Journal of Botany 52, 49–54.

Marrè, E. (1979). Fusicoccin: a tool in plant physiology. *Annual Review of Plant Physiology* **30**, 273–88.

Montalbini, P. & Buchanan, B. B. (1974). Effect of a rust infection on photophosphorylation by isolated chloroplasts. *Physiological Plant Pathology* **4**, 191–6.

Royle, D. J. (1976). Structural features of resistance to plant-diseases. In *Biochemical Aspects of Plant-Parasite Relationships*, ed. J. Friend & D. R. Threlfall, Phytochemical Society Symposium Series No. 13, pp. 161–93. London: Academic Press.

Safir, G. R. & Schneider, C. L. (1976). Diffusive resistances to two sugar beet cultivars in relation to their black root disease reaction. *Phytopathology* **66**, 277–80.

Smith, P. R. & Neales, T. F. (1977). Analysis of the effects of virus infection on the photosynthetic properties of peach leaves. *Australian Journal of Plant Physiology* **4**, 723–32.

Tipton, C. L., Mondal, M. M. & Benson, M. J. (1975). K^+-stimulated adenosine triphosphatase of maize roots: partial purification and inhibition by *Helminthosporium maydis* race T toxin. *Physiological Plant Pathology* **7**, 277–86.

Turner, N. C. & Graniti, A. (1976). Stomatal response of two almond varieties to fusicoccin. *Physiological Plant Pathology* **9**, 175–82.

Webster, J. (1970). *Introduction to Fungi*. London: Cambridge University Press.

Wheeler, B. E. J. (1969). *An Introduction to Plant Diseases*. London: John Wiley.

Wheeler, H. (1977). Ultrastructure of penetration by *Helminthosporium maydis*. *Physiological Plant Pathology* **11**, 171–8.

Yirgou, D. & Caldwell, R. M. (1963). Stomatal penetration of wheat seedlings by stem and leaf rust: effect of light and carbon dioxide. *Science* **141**, 272–3.

G. R. SQUIRE and C. R. BLACK
Stomatal behaviour in the field

Introduction

Understanding of stomatal behaviour in plants growing outdoors develops by the interdependent approaches of observation, experimentation and modelling. The first detailed information of stomatal responses to the environment was gained during the last ten years of the nineteenth and the first twenty years of the present century (Darwin, 1898; Lloyd, 1908; Loftfield, 1921; Sayre, 1926). Much useful information has been obtained since these pioneering investigations using several field techniques, such as liquid infiltration, surface impressions and portable viscous flow porometry, which are described earlier in this volume (Measurement of stomatal aperture and responses to stimuli: H. Meidner). These long established techniques remain useful and valid for purposes such as scheduling irrigation (Hack, 1978) or selecting for stomatal characteristics, particularly in the tropics where electronic equipment is often either too expensive or liable to breakdown.

Until ten to fifteen years ago, knowledge of stomatal behaviour of plants outdoors was based largely on studies which examined the infiltration of liquids into leaves (Hack, 1974, 1978). However, the phase of observation gained further momentum in the late 1960s with the development of porometers which measure the diffusive resistance of leaves to water vapour. The values obtained are predominantly influenced by stomatal aperture because the cuticles of most species are almost impermeable to water vapour. Consequently, stomatal responses may be followed readily by diffusion porometry. The stomatal resistances obtained may be used to estimate rates of transpiration from leaves provided boundary layer resistance, leaf temperature and atmospheric saturation deficit are also measured. Throughout this chapter, stomatal *conductance* rather than its reciprocal, resistance, will be used. The advantages of this convention for examining stomatal responses to the environment have been discussed by Burrows & Milthorpe (1976). The importance of stomatal conductance for diffusion of gases during photosynthesis and transpiration is discussed later in this volume (Stomatal control of transpiration and photosynthesis: P. G. Jarvis & J. I. L. Morison).

The experimental phase arose out of the need to explain observations, and to understand the effects on stomata of individual environmental variables. Much of the rest of this volume describes the current state of knowledge of experimental stomatal physiology. Although the chief environmental variables influencing stomata are now known, there are still considerable problems in understanding their exact mode of action on the stomata of plants growing outdoors. Consistent relations between stomatal conductance and individual environmental variables may be observed for short periods at one site, but cannot be applied universally since they may be modified by the age of the plants and the conditions experienced during growth. Consequently, predictions of conductance at one site are unreliable when they are based on relations established elsewhere, possibly under contrasting conditions.

Some of the inconsistencies between observations made on different stands of plants may be attributable to inaccuracies in the diffusion porometers used (e.g. Landsberg *et al.*, 1975). Although recent tests indicate that several porometers in current usage are reliable (Watts, 1977; Day, 1977; Black & Black, 1979), more rigorous analyses such as those described by Stigter and co-authors are required (Stigter, 1972; Stigter, Birnie & Lammers, 1973; Stigter & Lammers, 1974). However, even when measurements are made using accurate porometers, several major difficulties remain in interpreting observations and in producing models which predict the stomatal responses of plants growing out of doors. Some of these difficulties will be discussed in this paper.

Canopy conductance

A recent review by Körner, Scheel & Bauer (1979) illustrates the increased attention paid to field studies of stomatal behaviour over the last ten to fifteen years. Much of the financial support for these studies has been provided by institutions with a strong interest in agriculture and forestry, and much of the stimulus has come from scientists with a background in physics and engineering, whose aim was to understand and model the role of stomata in evaporation from canopies of plants. Many studies have concentrated on the stomatal response to dry conditions, because stomata are not usually the most important limitation to transpiration and photosynthesis in well-watered canopies. This is evident in the close relations between evaporation and the weather (Penman, 1948; Monteith, 1965; Van Bavel, 1967; Thom & Oliver, 1977), and between assimilation by plant stands and the amount of radiation they intercept (Monteith, 1972; Gallagher & Biscoe, 1978). For incomplete stands of vegetation, transpiration is determined largely by leaf area index (Stern, 1965; Ritchie & Burnett, 1971; Specht, 1972; Ritchie, 1973), provided the surface of the soil is dry. Even during drought, stomata are not always the major limitation to transpiration, because if drought is imposed slowly, as it usually is in the field, canopies may respond by reducing leaf area (Orshansky, 1954; Oppenheimer, 1960; Kozlowski, 1976), and large decreases in stomatal conductance may occur only as a last resort (Ritchie, 1973; Johns & Lazenby, 1977).

However, transpiration from canopies is always determined to some extent by both stomata and leaf area, two factors which have been combined in *canopy conductance* (Monteith, 1965). Estimates of canopy conductance from lysimetry (Van Bavel, 1967) or from synthesis of profiles of temperature, humidity and windspeed above a canopy (Monteith, Szeicz & Waggoner, 1965; Szeicz & Long, 1969) provide a good estimate of the influence of the environment on stomatal conductance. Canopy conduc-

Fig. 1. The relation between canopy conductance and intercepted radiation in a stand of winter wheat grown at Sutton Bonington in 1977. Each symbol is derived from measurements on about 30 leaves at several levels in the canopy. Mean saturation deficits during the measurements were: filled circles, 0–0.3 kPa; open circles, 0.3–0.6 kPa; triangles, above 0.6 kPa. The line indicates the relation under the most humid conditions.

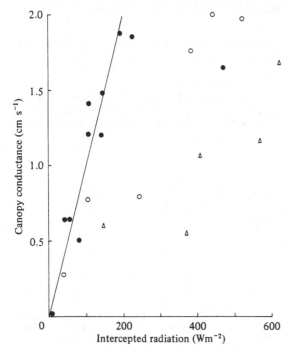

tance can also be estimated from measurements of leaf conductance with a porometer and leaf area index. In its simplest form, canopy conductance is the product of leaf conductance and leaf area index (Szeicz & Long, 1969; Szeicz, Van Bavel & Takami, 1973). However, because leaf conductance and stomatal behaviour vary considerably within stands, canopy conductance can be estimated more accurately by dividing the canopy into several distinct layers as described by Jarvis, James & Landsberg (1976) for a coniferous forest. Provided the stomatal conductance of each layer is substantially smaller than the aerodynamic conductance between layers, it is legitimate to calculate a canopy conductance by summing the layer conductances. Canopy conductance is therefore the sum of the leaf conductance in each layer weighted by the corresponding leaf area index (Table 1). Estimates of leaf area index, or alternatively, green area index in species where organs such as leaf-sheaths form an appreciable portion of the transpiring surface, are therefore necessary for a complete understanding of stomatal responses in the field.

The assumption that stomatal conductances are smaller than aerodynamic conductances is usually valid near the top of a canopy where turbulent mixing is most intense, and remains valid in the layers of foliage which contribute most of the water vapour. Leaves near the bottom of complete canopies contribute relatively little to transpiration because stomatal conductances are usually small. In the stand of wheat (*Triticum aestivum* L. cv

Fig. 2. The relation between stomatal conductance and irradiance. (*a*) adaxial, (*b*) abaxial surfaces of sunlit leaves of pearl millet grown in India. Conductances are means of measurements on five leaves made over 30 min; standard errors are about twice the size of the symbols. Irradiance is the average for the period. Open circles, irrigated; filled circles, unirrigated. The line in (*a*) is the linear regression of all the symbols; the dashed line in (*b*) separates most irrigated and unirrigated symbols. See text for an explanation of the triangles in (*b*).

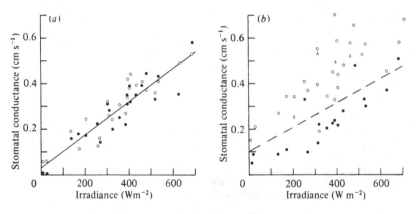

Table 1. *An example of the calculation of canopy conductance* ($g_g = 0.88$ cm s^{-1}) *of a stand of pearl millet grown in India*

Part of canopy	g_l[a] (cm s^{-1})	GAI[b]	$g_l \times$ GAI (cm s^{-1})
Leaves:			
0–30 cm above soil	0.13 (0.030)[c]	0.29 (0.10)	0.04
30–60 cm	0.55(0.12)	0.77(0.055)	0.42
Above 60 cm	1.08(0.075)	0.35(0.075)	0.38
Sheaths[d]	0.19(0.096)	0.21(0.017)	0.04
Total		1.62	0.88

[a] Values of leaf conductance (adaxial plus abaxial) are means of five measurements on each surface in each part of the canopy. The series of measurements lasted 30 min.

[b] Green area indices are the means of four samples, each of 90 plants.

[c] Figures in brackets are standard errors.

[d] For consistency with leaves, the sheath area index is derived from measurements of planar area, and sheath conductance is twice the conductance measured on one side of a sheath.

Fig. 3. Profiles of leaf conductance (adaxial plus abaxial) in a stand of winter wheat on 25 May (filled circles), 21 June (open circles), and 15 July (triangles) 1977. Single standard errors are shown. Further details are given in Table 2.

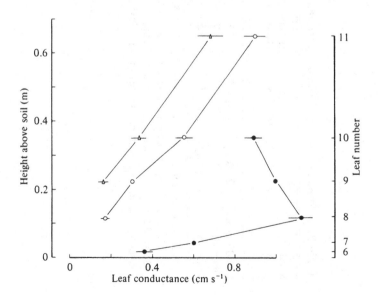

Table 2. *Information on microclimate, area index, and canopy conductance relevant to Fig. 3*

Symbol on Fig.3	Date	Time (h, GMT)	Irradiance (W m^{-2})	Intercepted radiation[a] (W m^{-2})	Saturation deficit[b] (kPa)	Air temperature[b] (°C)	Area index Leaves	Sheaths	Canopy conductance[c] (cm s^{-1})
Filled circles	25 May	14–15	730	382	0.58	17.6	1.9	0.3	1.78
Open circles	21 June	14–15	675	623	0.65	17.4	2.9	1.4	1.69
Triangle	15 July	13–14	656	629	0.82	18.3	2.1	1.5	1.72

[a] Average value for the hour measured with tube solarimeters.

[b] Average value for the hour measured with an aspirated psychrometer at 1.5 m.

[c] Derived from measurements of conductance and area index on all leaves shown on Fig. 3 and sheaths. Mean conductances for each leaf layer were obtained from measurements on adaxial and abaxial surfaces of, usually, six leaves.

Maris Huntsman) described later in this paper (Figs. 1, 3), the top two leaves and their sheaths contributed over 70% to total canopy conductance. For the more open stand of pearl millet (*Pennisetum typhoides* S. and H.) (Table 1; Fig. 2), where light interception was incomplete, the bottom layer of leaves, 0–30 cm above the soil, contributed only 10% to canopy conductance. Further information and discussion on the physical aspects of evaporation and canopy conductance are contained in a recent review by Shuttleworth (1979).

Most studies of canopy conductance have been micrometeorological, and where porometers have been used, it has been to distinguish between water evaporated from the soil and water transpired from plants. The possibility of using canopy conductance in physiological studies has hardly been explored. Evidence summarized in the next section of this chapter shows that measurements of leaf conductance in a single layer or part of a canopy may give an inadequate description of the stomatal response of the stand as a whole. A more complete description of stomatal responses to the environment requires detailed measurements of profiles of conductance, which are difficult to use for comparisons between stands if the stands differ in height or in leaf area index. The combination of data from profiles into a single value of a canopy conductance, which also takes account of leaf area, may prove to be a useful method of analysis in physiological studies. For example, 22 profiles, each similar to the ones in Fig. 3, are summarized in Fig. 1. The result shows relations between conductance and two environmental variables, sunlight and humidity, which were not detected by examination of either single layers or of profiles.

The variable environment

The outdoor environment, as perceived by plants, is rarely constant for more than a few minutes, and variations in sunlight, temperature and humidity are often closely linked. Changes in the amount of water in the soil over periods of days or weeks may interact with the other variables and modify their influence.

It is difficult therefore to isolate the effect of one variable if the stomata respond to several at the same time. For example, it may be impossible to obtain the true relation between conductance and radiation purely from field observations. This difficulty is illustrated for a stand of winter wheat in Fig. 1 (cf. Jarvis, 1976). Canopy conductance was calculated from measurements of leaf conductance and leaf area index for at least four layers in the canopy on several days in May and June. Each point on the figure is derived from measurements on about 30 leaves made during the course of one hour;

values of radiation and saturation deficit are averages for this period. Canopy conductance is plotted in relation to intercepted solar radiation rather than irradiance, to take account of changes in leaf area index from day to day. Conductance rose sharply with intercepted radiation but apparently reached a maximum value at $200 \, W \, m^{-2}$. It is impossible to determine whether this value of conductance represents a true maximum, because atmospheric saturation deficit usually exceeded 0.3 kPa whenever intercepted radiation rose above $200 \, W \, m^{-2}$. The stomata of wheat are responsive to changes in ambient humidity and show a partial closure in dry air. The true relation between conductance and radiation for this stand of wheat would have been detectable only if other factors had been constant.

Rapid fluctuations in the outdoor environment, and rapid stomatal responses, also hinder the accurate measurement and interpretation of stomatal conductance. An essential feature of porometry is that the stomatal response to the presence of the porometer cup is slower than the time required to get a stable measurement. The stomata of many species, such as the wheat, tobacco (*Nicotiana tabacum* L.), groundnut (*Arachis hypogaea* L.) and millet plants discussed in this paper do not respond within one to two minutes of enclosure, but other species may be more sensitive. Porometers which modify the ambient conditions only slightly are an advantage when examining the conductance of species whose stomata react quickly (within 15 s) to a change in the environment.

The rate of stomatal response to fluctuations in the environment also influences the degree of scatter obtained between conductance and particular environmental variables. For example, instantaneous measurements of conductance and irradiance may be poorly correlated if the stomatal response is slower than the rate of change of irradiance. In many species, including pearl millet, wheat, groundnut and tobacco, stomatal conductance requires one or two minutes to respond appreciably to changes in irradiance, and probably 30 to 60 min to respond completely to variations in potential transpiration. Similar, or slightly longer, response-times of conductance to irradiance have been observed for several woody species (Woods & Turner, 1971; Davies & Kozlowski, 1974); other species, such as Sitka spruce (*Picea sitchensis* (Bong.) Carr.) are much slower in response to environmental change as described later (see Stomatal control of transpiration and photosynthesis: P. G. Jarvis & J. I. L. Morison). One method which may be used to minimize this problem is to use mean values of conductance and environmental variables which have been measured repeatedly over perhaps 30 or 60 min. However, this approach may give misleading results if the environment varies appreciably during the measurements.

The variable plant

At any instant, stomatal conductance varies throughout a canopy: within leaf surfaces, between upper and lower surfaces, between leaves and between plants. At any point in a canopy, conductance varies with time even under constant conditions, owing to the effect of age. Therefore the observed response of conductance to any environmental variable will depend on where and when conductance is measured.

Spatial variation

Stomatal conductance may vary within leaves by a factor of two or three within a distance of several centimetres. Variations are usually most pronounced in species with large leaves, or leaves whose hydraulic systems distribute water inefficiently (Rawlins, 1963). In the study from which Fig. 1 is taken, the variation in stomatal conductance within wheat leaves was not systematic, and when the data for several leaves were normalized, mean conductance was similar at all points along the leaf except near the apex, where conductance increased slightly. There is little information on whether these local variations result from changes in stomatal frequency or size, out-of-phase oscillations, or inherent differences in the stomatal response to the environment. However, during water stress, the variations within the leaves of grasses may be large and systematic with conductance decreasing towards the apex. This type of variation may be related to the large gradients of water potential which may develop within leaves when transpiration is fast (Rawlins, 1963; Campbell, Zollinger & Taylor, 1966).

Stomatal conductance may differ between leaf surfaces because of differences in stomatal frequency. An example is *Vicia faba* L. whose adaxial and abaxial conductances under favourable environmental conditions differ to an extent related to stomatal frequency (Black & Black, 1979). In other species with similar numbers of stomata on both surfaces, conductance may differ between surfaces because of differences in microenvironment or response to the environment. Differences between adaxial and abaxial surfaces in stomatal responses to light or water stress have been reported by several investigators, including Kanemasu & Tanner (1969), Turner (1970), and Sharpe (1973). Porometers which permit the conductances of both surfaces to be measured separately are invaluable in such instances. For example, the relation between incident irradiance (on a plane parallel to the surface of the canopy) and conductance of abaxial and adaxial surfaces of leaves of pearl millet at the top of a canopy is shown in Fig. 2. Measurements were made in Central India on irrigated and unirrigated plants during a period of rainless weather when reserves of moisture in the soil were being depleted by the plants (Squire, 1979). For the upper (adaxial) surface, the relation was similar in both irrigated and unirrigated plants throughout the period of measurements, lasting about fifteen days. This relation is another example of a linear, rather than hyperbolic, response of conductance to sunlight (cf. Biscoe, Cohen & Wallace, 1976; Denmead & Millar, 1976). During these measurements atmospheric saturation deficit varied between 0.4 and 3.0 kPa, and air temperature between 23 and 31 °C. This implies one of two things: either that conductance responded only to irradiance and was insensitive to other factors; or that the net effect of other factors on conductance was negligible. In contrast there was a clear separation between irrigated and unirrigated plants for the lower surface, implying that the stomata on the upper and lower surfaces were responding differently to the amount of water in the soil. The physiological and anatomical bases for this type of response are not known, and would be worth further study. The reduction of abaxial conductance on unirrigated plants was reversible, as is demonstrated by the triangular symbols which represent values obtained on the day following an isolated shower of rain during an otherwise dry period. In this example, a porometer which measured the conductance of both leaf surfaces at the same time would have failed to separate the contrasting stomatal responses, and would have resulted in a light-response curve which did not accurately describe the response of millet stomata to light.

In the same study, stomatal conductance of the middle (30–60 cm above the ground) and upper (60–90 cm above the ground) layers of leaves responded differently to the environment. In particular, when the rate of potential transpiration increased, conductance was reduced relatively more in the middle than in the upper part of the canopy – a response to water stress observed in stands of other species (Teare & Kanemasu, 1972; Denmead & Millar, 1976). This response is also consistent with experiments in the laboratory on whole plants, where stomata closed earlier on lower (older) leaves during periods of water stress (Jordan, Brown & Thomas, 1975), possibly because younger leaves at the top of the canopy may be preferentially supplied with water. However, the largest differences in conductance between leaves in canopies are usually attributed to variation in irradiance caused by shading. For example, Denmead & Millar (1976) and Turner & Incoll (1971) observed for wheat and sorghum, respectively, that a constant relation between conductance and irradiance held for all leaves regardless of their position in the canopy. Similar relations will not hold for other species if the stomatal response changes with age, as was observed, for example, by Turner & Incoll (1971) for tobacco.

Investigators of stomatal behaviour on plants growing outdoors often encounter large differences in conductance between apparently similar adjacent plants, but there have been few detailed studies of the relative contribution of the genetic or environmental components of this variation (Jones, 1974). In practice, a satisfactory mean value of conductance can usually be obtained within about five minutes, from measurements on five to ten leaves. Methods for screening stands and results of selection for stomatal characteristics are reviewed by Jones (1979).

Temporal variation

Evidence such as that in Fig. 2a illustrates that the stomatal response to the environment of some epidermes can remain constant for several days or weeks, especially during vegetative growth. Over longer periods, stomatal responses may change with age (Turner & Heichel, 1977; Jones, 1979; Giurgevich & Dunn, 1979), in response to specific developmental stimuli such as fruiting (Hansen, 1971), or as a result of after-effects of environmental stresses (Neilson & Jarvis, 1975; Fahey, 1979; see Begg & Turner (1976) and W. J. Davies *et al.* in this volume for references to water stress). Often, in practice, it is difficult to separate the effects of environmental and endogenous factors, because ageing occurs on a similar time-scale to seasonal change in environment and in leaf area index. To illustrate this point, profiles of leaf conductance for a stand of winter wheat grown at Sutton Bonington, England, are shown in Fig. 3. The appropriate leaf area indices, canopy conductances and micrometeorological information are given in Table 2. On the earliest date shown, leaf conductance was larger on the fully expanded leaves (8 and 9) than on the young expanding leaves which were in full sunlight. The profiles for the other two dates are characteristic of closed canopies, where conductance was largest at the top of the canopy in leaves receiving full sunlight. Environmental conditions during the three periods were similar, although the slightly larger saturation deficit on the last date may have been responsible for the slightly smaller conductances. The values of canopy conductance shown in Table 2 were obtained by summing the products of leaf conductance and area index for each leaf and the sheaths. Canopy conductance was similar on all three occasions, despite an almost twofold change in green area (leaves plus sheaths), because of compensatory changes in leaf conductance. The reduction in the conductance of most leaves between 25 May and 21 June could be attributed to ageing, but may alternatively be a result of the increase in the area of evaporating surface.

Any change in leaf area is likely to alter the demand for water imposed by the foliage on the roots. The possible interdependence of stomatal conductance and leaf area is illustrated by the following two examples. The effect of a progressive reduction of transpiring leaf area on transpiration from cotton (*Gossypium hirsutum* L.) plants grown in pots is shown in Fig. 4. The transpiring area was reduced by covering leaves successively from the base of the stem to avoid shading of lower leaves which normally transpired at about one-third the rate of younger, fully expanded leaves. The rate of transpiration per plant remained unchanged until over 60% of the leaf area had been covered, owing to a simultaneous compensatory increase in the transpiration rate of the remaining leaves. Although no measurements were made, the progressive increase in the transpiration rate of the uncovered foliage implies a similar increase in stomatal conductance. A comparable response to artificial modifications of leaf area was observed by Kelley (1932).

In the second example (Fig. 5), using pearl millet grown with an un-restricted root system in the soil within a glasshouse, plants with a wide range of exposed leaf areas were obtained by temporarily covering some of the leaves with aluminium foil. Mean leaf conductance was relatively small in plants with most leaves uncovered because of depletion of moisture reserves in the soil at this stage of growth. However, reduction of the transpiring area resulted in an increase in the conductance of the uncovered leaves. Calculation of the appropriate canopy conductances (Fig. 5*b*) demonstrates that the decrease in mean leaf conductance was insufficient to compensate fully for the increasing transpiring area. However, the broken line shows the values of canopy conductance predicted in the absence of an interaction

Fig. 4. The effect of decreasing leaf area on transpiration in cotton. Circles, the rate of transpiration from the whole plant; triangles, the rate of transpiration per unit area of transpiring surface; squares, area of transpiring surface.

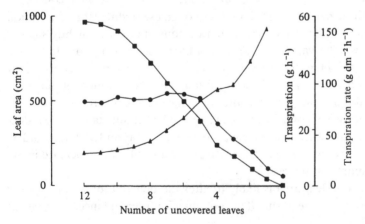

between the conductance of individual leaves and the area of transpiring foliage.

The importance of changes in leaf area in natural communities is often obscured by simultaneous changes in the size of the root system. Despite this complication, there is evidence, cited earlier, that the stomatal response to drought may sometimes be small, and that alterations in leaf area may be of major importance. This point is illustrated by Table 3 which contains values for leaf and canopy conductance and leaf area index of two tobacco varieties grown with unrestricted root systems in glasshouses, either in a drying soil or in soil maintained near field capacity (Pearson, 1980). The responses of the two varieties to drought form a clear contrast. Speight G13, a commercial

Fig. 5. (*a*) The effect of reducing leaf area on the leaf conductance (adaxial plus abaxial) of sunlit leaves of pearl millet. (*b*) The same data expressed as canopy conductance. The text gives an explanation of the dashed line. Adapted from Black & Squire (1979).

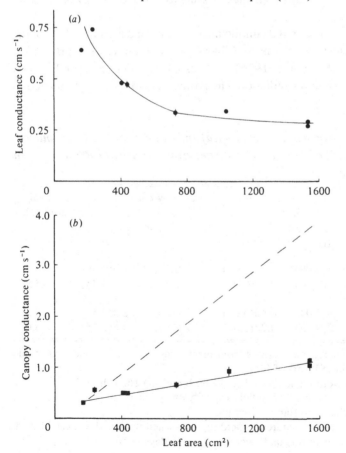

American variety of *Nicotiana tabacum*, responded to drought by reducing leaf area, and had similar leaf conductances in both treatments. Hishi, a Jordanian member of *N. rustica* L. normally grown with a limited water supply, showed a reduction in both leaf area and conductance under drought, resulting in a much greater decrease in canopy conductance than observed in Speight G13.

Experimentation and modelling

The aim of modelling stomatal responses is to enable predictions of conductance to be made from a knowledge of the species and variety of plant, the history of growth, and current environment. Models have diverse applications in hydrology, agriculture, and forestry for predicting evaporation (Black, Tanner & Gardner, 1970; Ritchie, 1972; Kanemasu, Stone & Powers, 1976; Shuttleworth, 1979), and also in experimental investigations of stomatal responses and their mechanisms (Schulze *et al.*, 1974; Cowan, 1977).

Several models of stomatal conductance have been developed. One of the simplest and most widely used (Shawcroft, Lemon & Stewart, 1973; Sinclair, Murphy & Knoerr, 1976) assumes a hyperbolic relation between stomatal resistance and irradiance. The parameters in the equation need to

Table 3. *The influence of drought on leaf and canopy conductance, and leaf area index of stands of two tobacco varieties grown with their roots in the earth inside glasshouses*

Variety and treatment	Conductance of sunlit leaves (cm s^{-1})		Canopy conductance[a] (cm s^{-1})		Leaf area index
	(i)[b]	(ii)	(i)	(ii)	
Speight G13					
Irrigated[c]	0.98(0.09)[d]	0.39(0.01)	4.1(0.30)	1.9(0.30)	6.4
Unirrigated[e]	1.03(0.10)	0.39(0.01)	2.4(0.30)	1.2(0.35)	3.9
Hishi					
Irrigated	0.98(0.04)	0.50(0.01)	3.9(0.10)	2.3(0.50)	5.8
Unirrigated	0.72(0.05)	0.28(0.01)	1.1(0.35)	0.6(0.35)	3.0

[a] Canopy conductance was derived from measurements of leaf conductance and area index of three layers of foliage.
[b] Measurements were made around (i) 11.00 h and (ii) 15.00 h GMT.
[c] Irrigated plots were watered to field capacity each day.
[d] Figures in brackets are standard errors.
[e] Unirrigated plots were watered to field capacity when the stand was established. The measurements in this table were made sixty days later.

be derived from observations, and have to be changed depending on the severity of water stress. In a model described by Schulze *et al.* (1974), the relations between conductance and two environmental variables, temperature and humidity, were derived, not from observations, but from experiments in which leaves of *Prunus armeniaca* L. growing in the Negev were enclosed in controlled environment cuvettes. The relations were then used to predict, with notable success, the diurnal course of conductance from measurements of temperature and humidity. In a third example (Jarvis, 1976), the effects on conductance of five environmental variables (light, temperature, water potential, saturation deficit and CO_2-concentration) can be combined in one equation. The value of this approach for analysing scatter diagrams of conductance and environmental variables has been demonstrated by Callander & Woodhead (1981) in their analysis of diurnal and seasonal trends of canopy conductance of a tea plantation in Kenya. In the fourth and final example (Cowan, 1977; Cowan & Farquhar, 1977), the model predicts values of stomatal conductance which would enable plants to lose as little water as possible for the carbon they fix in photosynthesis. In the model, stomata respond to the environment in such a way that the resulting changes in evaporation and assimilation bear a constant numerical relation to each other (constant, at least for a period of one or several days), with the result that the concentration of carbon dioxide in the intercellular space remains constant.

In these and other models the parameters used in the equations have to be derived either from observations made over a wide range of conditions (Jarvis, 1976), or from experiments in controlled environments. However, it is difficult to obtain the relevant terms from observations alone. Although idealized light-response curves have been derived from measurements on several species (see reviews by Turner (1974) and Burrows & Milthorpe (1976)), it is doubtful if these curves can be applied in all environments, even when soil water is not limiting. Examples are the modifying influence of other factors on the response of conductance to light in winter wheat (Fig. 1) and in Sitka spruce (Jarvis, 1976; Watts, 1977).

Some form of experimentation under controlled conditions is clearly essential to establish the dimension and nature of the variables included in models. But because different parts of plants exhibit different responses to the environment, and because the nature of the response may depend on age, history, leaf area, and amount and distribution of roots, it is doubtful whether experiments in laboratories or growth rooms, with plants grown in small containers with restricted roots, can provide more than the general form of the response. The major problems involved in interpreting field observations from the results of laboratory experiments, particularly with respect to water relations, have been discussed by Begg & Turner (1976) in a review of the effects on crops of water deficits, and by Taylor & Klepper (1978) who considered the absorption of water by roots. Recreation of realistic outdoor environments in growth rooms presents major difficulties (Van Bavel, 1973; Downs & Hellmers, 1975; Raper & Downs, 1976). For instance, although in some species stomatal conductance reaches a maximum at solar irradiances of $150-200 \, W \, m^{-2}$, in other species conductance continues to increase with irradiance up to $800-1000 \, W \, m^{-2}$ (Turner, 1974; Biscoe, Cohen & Wallace, 1976; Denmead & Millar, 1976). These irradiances are five to ten times larger than those commonly used in laboratories or growth rooms. Furthermore, the spectral quality of sunlight cannot easily be reproduced in growth chambers, since the spectral energy distribution within the visible wavelengths of artificial sources of light differs greatly from sunlight, and the proportion of wavelengths above 700 nm is much reduced. However, the main contrast between field and growth room is that, in the field, drought is usually imposed sufficiently slowly to allow plants to adapt by adjusting their osmotic potential, by reducing leaf area or by increasing the growth of the root system. Several examples of different stomatal responses to water deficits in the field and in the laboratory are quoted by Begg & Turner (1976).

More realistic experimental systems using plants with unrestricted roots, growing in natural or near-natural environments can be grouped into three broad categories: (i) cuvettes enclosing branches, individual leaves, or parts of leaves of plants growing out of doors; (ii) larger chambers enclosing individual plants or groups of plants growing out of doors; and (iii) glasshouses enclosing small stands of plants.

Cuvettes enclosing parts of plants are extremely useful for rapid measurements of photosynthesis and transpiration, from which stomatal conductance is calculated, under different environmental conditions (Mooney et al., 1971; Schulze et al., 1972; Nobel, 1976). Results of a productive series of experiments using this method have been reviewed by Hall, Schulze & Lange (1976). It has the disadvantage that the enclosed tissue is subjected to a different environment from the rest of the plant. Consequently, misleading results may be obtained if the conductance of the enclosed leaf is influenced by the environment of the other leaves, or by transpiration from the plant as a whole, as was suggested earlier. For instance, Rawson, Begg & Woodward (1977) found that the conductance of barley leaves responded to changes in atmospheric humidity when the entire plant was subjected to the treatment, but not when single leaves enclosed in a cuvette were treated.

Larger chambers or enclosures (Jarvis, Eckardt, Koch & Čatský, 1971;

Leafe, 1972; Schulze, 1972; Louwerse & Eikhoudt, 1975) overcome this problem but cannot detect differences between leaves unless used in conjunction with a diffusion porometer. Nor do chambers and most cuvettes permit the separation of adaxial and abaxial conductances, except by porometry. Chambers are ideally suited for use with isolated plants or widely spaced crops, but they interfere with the microclimate in dense stands. Controlled environment glasshouses overcome this problem but again require the use of porometry to characterize the stomatal response of individual leaves. Glasshouses also modify the light-environment more severely than cuvettes, particularly with respect to the quantity and quality of incident radiation, owing to the permanence of the enclosure.

The data for pearl millet in Fig. 5 provide an example of the type of work which can be undertaken in controlled environment glasshouses. The plants were grown with unrestricted root systems, under conditions of temperature and humidity intended to reproduce those of the natural habitat as closely as possible. Air temperature was varied diurnally, sinusoidally, with an amplitude of 5 °C around different means between 19 °C and 31 °C. Maximum irradiance at the canopy surface exceeded 600 W m^{-2} on bright days in summer. Although this value is lower than maximum values in the tropics, the total radiation received by the plants in a day was comparable because of the long summer days in temperate regions. Thus the plants were grown under varying, but precisely known, conditions approximating the natural

Fig. 6. Response of leaf conductance (adaxial plus abaxial) of groundnut plants, grown with their roots in the earth, to atmospheric saturation deficit in a controlled-temperature glasshouse. Air temperature was 33.5 ± 2 °C throughout, and irradiance incident on the leaves was 580 ± 20 W m^{-2} from 10.00 h to 15.00 h decreasing to 450 W m^{-2} between 15.00 h and the end of the measurements.

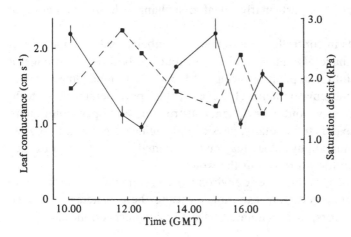

environment. Periods when conditions were controlled or varied for experiments could be inserted into the overall pattern of diurnal variation. For instance, experiments at constant temperature demonstrated that the stomata of pearl millet and groundnut are highly responsive to variations in the atmospheric saturation deficit (Fig. 6; Black & Squire, 1979). The stomatal responses were of comparable size throughout the day in groundnut, but were reduced or absent in the morning in millet. This inconsistency in the stomatal behaviour of pearl millet implies that reserves of water in the plant and adjacent to the roots may be sufficient to cope with variations in evaporative demand during the early part of the day, but not later when reserves would have been partially depleted. It is unlikely that effects such as this would be detected using cuvettes attached to single leaves, or would be observed at all on plants with restricted root systems.

Conclusions

The invention and subsequent refinement of diffusive resistance porometers provided a major stimulus to the study of stomatal behaviour in the field. The ease and quickness with which conductance can now be measured has led to the accumulation of a mass of information which is sometimes inconsistent, and often difficult to analyse. Because stomatal responses to the environment vary so much even within stands of a single species, measurements made over one or two days on a small part of a canopy are likely to lead only to further confusion. There is still the need for detailed observations on all parts of the canopy for extended periods of weeks or months, particularly if additional information on the environment and on the state of the plant is available. The interpretation of stomatal responses in terms of canopy conductance may be useful when dealing with stands of vegetation, particularly if leaf area changes during the period of study.

Experiments in controlled environment chambers, enclosing parts of or whole plants, have also made major contributions both to knowledge of stomatal physiology of plants growing outdoors, and to modelling stomatal responses to the environment. However, unlike porometers, these techniques are expensive and their design, construction, and operation require considerable expertise in botany, physics and engineering. Future progress in outdoor stomatal physiology may be determined by the number of such systems in operation throughout the world.

The effects of the atmospheric environment on stomatal conductance are now well documented from experiments using these systems. Although atmospheric factors, such as humidity, have large effects on the stomata of some species, to others, including many crops, they are less important. For example, the values of leaf conductance observed in bright sunlight in the glasshouse experiment mentioned at the end of the last section, could not be forced much below $1 \, cm \, s^{-1}$ for groundnut and $0.6-0.7 \, cm \, s^{-1}$ for millet, even at saturation deficits of $3.0 \, kPa$, provided there was plenty of water in the soil. These values are comparable to the maximum conductances observed at similar saturation deficits in pearl millet grown outdoors in India (Squire, 1979), and are large enough to allow rapid photosynthesis and transpiration. In these species, stomatal conductances low enough to restrict transpiration and photosynthesis severely are more likely to result from shortage of water in the soil than from dry air. Relatively little is known of the quantitative relations between canopy conductance and the supply of water, determined by the distribution, density and activity of roots in relation to the distribution of available water. These relations are particularly relevant in arid or semi-arid lands where soil water potential may increase sharply from $-1.5 \, MPa$ or lower at the surface, to values exceeding $-0.1 \, MPa$ at a depth of one to two metres. For a better understanding of stomatal responses to drought, stomatal investigations need to be accompanied by detailed studies of rooting and absorption of water from the soil.

References

Begg, J. E. & Turner, N. C. (1976). Crop water deficits. *Advances in Agronomy* **28**, 161–217.

Biscoe, P. V., Cohen, Y. & Wallace, J. S. (1976). Daily and seasonal changes of water potential in cereals. *Philosophical Transactions of the Royal Society, London* Series B **273**, 565–80.

Black, C. R. & Black, V. J. (1979). The effect of low concentrations of sulphur dioxide on stomatal conductance and epidermal cell survival in field bean (*Vicia faba* L.). *Journal of Experimental Botany* **30**, 291–8.

Black, C. R. & Squire, G. R. (1979). Effects of atmospheric saturation deficit on the stomatal conductance of pearl millet (*Pennisetum typhoides* S. and H.) and groundnut (*Arachis hypogaea* L.). *Journal of Experimental Botany* **30**, 935–45.

Black, T. A., Tanner, C. B. & Gardner, W. R. (1970). Evaporation from a snap bean crop. *Agronomy Journal* **62**, 66–9.

Burrows, F. J. & Milthorpe, F. L. (1976). Stomatal conductance in the control of gas exchange. In *Water Deficits and Plant Growth*, Volume IV, ed. T. T. Kozlowski, pp. 103–52. New York: Academic Press.

Callander, B. A. & Woodhead, T. (1981). Canopy conductance of estate tea in Kenya. *Agricultural Meteorology* **22** (In press).

Campbell, G. S., Zollinger, W. D. & Taylor, S. A. (1966). Sample changer for thermocouple psychrometers: construction and some applications. *Agronomy Journal* **58**, 315–18.

Cowan, I. R. (1977). Stomatal behaviour and environment. In *Advances in Botanical Research* **4**, ed. R. D. Preston & H. W. Woolhouse, pp. 117–228. London: Academic Press.

Cowan, I. R. & Farquar, G. D. (1977). Stomatal function in relation to leaf metabolism and environment. In *Integration of Activity in the Higher Plant. Symposium of the Society for Experimental Biology* **31**, 471–505.

Darwin, F. (1898). Observations on stomata. *Philosophical Transactions of The Royal Society* Series B **190**, 531–621.

Davies, W. J. & Kozlowski, T. T. (1974). Stomatal responses of five woody angiosperms to light and humidity. *Canadian Journal of Botany* **52**, 1525–34.

Day, W. (1977). A direct reading continuous flow porometer. *Agricultural Meteorology* **18**, 81–9.

Denmead, O. T. & Millar, B. D. (1976). Field studies of the conductances of wheat leaves and transpiration. *Agronomy Journal* **68**, 307–11.

Downs, R. J. & Hellmers, H. (1975). *Environment and the Experimental Control of Plant Growth.* London: Academic Press.

Fahey, T. J. (1979). The effect of night frost on the transpiration of *Pinus contorta* ssp. *latifolia. Oecologia Plantarum* **14**, 483–90.

Gallagher, J. N. & Biscoe, P. V. (1978). Radiation absorption, growth and yield of cereals. *Journal of Agricultural Science* **91**, 47–60.

Giurgevich, J. R. & Dunn, E. L. (1979). Seasonal patterns of CO_2 and water vapour exchange of the tall and short height forms of *Spartina alterniflora* Loisel in Georgia salt marsh. *Oecologia* **43**, 139–56.

Hack, H. R. B. (1974). The selection of an infiltration technique for estimating the degree of stomatal opening of leaves in field crops in the Sudan and a discussion of the mechanism which controls the entry of test liquids. *Annals of Botany* **38**, 93–114.

Hack, H. R. B. (1978). Stomatal infiltration in irrigation experiments on cotton, grain sorghum, groundnuts, kenaf, sesame and wheat. *Annals of Botany* **42**, 509–47.

Hall, A. E., Schulze, E. D. & Lange, O. L. (1976). Current perspectives of steady-state stomatal responses to the environment In *Water and Plant Life. Problems and Modern Approaches,* ed. O. L. Lange, L. Kappen, & E. D. Schulze, pp. 169–88. Berlin: Springer-Verlag.

Hansen, P. (1971). The effect of fruiting upon transpiration rate and stomatal opening in apple trees. *Physiologia Plantarum* **25**, 81–3.

Jarvis, P. G. (1976). The interpretation of the variation in leaf water potential and stomatal conductance found in canopies in the field. *Philosophical Transactions of The Royal Society, London* Series B **273**, 593–610.

Jarvis, P. G., Eckardt, F. E., Koch, W. & Čatský, J. (1971). Examples of assimilation chambers in current use. In *Plant Photosynthetic Production: Manual of Methods,* ed. Z. Šesták, J. Čatský, & P. G. Jarvis, pp. 84–110. The Hague: Dr. W. Junk.

Jarvis, P. G., James, G. B. & Landsberg, J. J. (1976). Coniferous forests. In *Vegetation and the Atmosphere, Vol. 2, Case Studies,* ed. J. L. Monteith, pp. 171–240. New York: Academic Press.

Johns, G. G., & Lazenby, A. (1977). Defoliation, leaf area index and the water use of four temperate pasture species under irrigated and dryland conditions. *Australian Journal of Agricultural Research* **24**, 783–95.

Jones, H. G. (1974). Assessment of stomatal control of plant water status. *New Phytologist* **73**, 851–9.

Jones, H. G. (1979). Stomatal behaviour and breeding for drought tolerance. In *Stress Physiology in Crop Plants*, ed. H. Mussell & R. C. Staples, pp. 407–28. New York: John Wiley & Sons.

Jordan, W. R., Brown, K. W. & Thomas, J. C. (1975). Leaf age as a determinant in stomatal control of water loss from cotton during water stress. *Plant Physiology* **56**, 595–9.

Kanemasu, E. T., Stone, L. R. & Powers, W. L. (1976). Evapotranspiration model tested for soybean and sorghum. *Agronomy Journal* **68**, 569–72.

Kanemasu, E. T. & Tanner, C. B. (1969). Stomatal diffusive resistance of snap beans. I. Influence of leaf-water potential. *Plant Physiology* **44**, 1547–52.

Kelley, U. W. (1932). The effect of pruning of excised shoots on the transpiration rate of some deciduous fruit species. *Proceedings of the American Society for Horticultural Science* **29**, 71–3.

Körner, C., Scheel, J. A. & Bauer, H. (1979). Maximum leaf diffusive conductance in vascular plants. *Photosynthetica* **13**, 45–82.

Kozlowski, T. T. (1976). Water supply and leaf shedding. In *Water Deficits and Plant Growth*, Vol. IV, ed. T. T. Kozlowski, pp. 191–231. New York: Academic Press.

Landsberg, J. J., Beadle, C. L., Biscoe, P. V., Butler, D. R., Davidson, B., Incoll, L. D., James, G. B., Jarvis, P. G., Martin, P. J., Neilson, R. E., Powell, D. B. B., Slack, E. M., Thorpe, M. R., Turner, N. C., Warrit, B., & Watts, W. R. (1975). Diurnal energy, water and CO_2 exchanges in an apple *(Malus pumila)* orchard. *Journal of Applied Ecology* **12**, 659–84.

Leafe, E. L. (1972). Micro-environment, carbon dioxide exchange and growth in grass swards. In *Crop Processes in Controlled Environments*, ed. A. R. Rees, K. E. Cockshull, D. W. Hand & R. G. Hurd, pp. 157–74. London and New York: Academic Press.

Lloyd, F. E. (1908). The physiology of stomata. *Publication of the Carnegie Institution, Washington* **82**, 1–142.

Loftfield, J. V. G. (1921). The behaviour of stomata. *Publication of the Carnegie Institution, Washington*, **314**, 1–104.

Louwerse, W., & Eikhoudt, J. W. (1975). A mobile laboratory for measuring photosynthesis, respiration and transpiration of field crops. *Photosynthetica* **9**, 31–3.

Monteith, J. L. (1965). Evaporation and environment. In *The State and Movement of Water in Living Organisms*, ed. G. E. Fogg. *Symposium of the Society for Experimental Biology* **19**, 205–34.

Monteith, J. L. (1972). Solar radiation and productivity in tropical ecosystems. *Journal of Applied Ecology* **9**, 747–66.

Monteith, J. L., Szeicz, G. & Waggoner, P. E. (1965). The measurement and control of stomatal resistance in the field. *Journal of Applied Ecology* **2**, 345–55.

Mooney, J. A., Dunn, E. L., Harrison, A. T., Morrow, P. A., Bartholomew, B., & Hays, R. L. (1971). A mobile laboratory for gas exchange measurements. *Photosynthetica* **5**, 128–32.

Neilson, R. E. & Jarvis, P. G. (1975). Photosynthesis in Sitka spruce *(Picea sitchensis* (Bong.) Carr.). VI. Response to temperature. *Journal of Applied Ecology* **12**, 879–91.

Nobel, P. S. (1976). Water relations and photosynthesis of a desert CAM plant *Agave deserti*. *Plant Physiology* **58**, 576–82.

244   G. R. SQUIRE & C. R. BLACK

Oppenheimer, H. R. (1960). Adaptation to drought: xerophytism. In
    *Plant Water Relations in Arid and Semi-arid Conditions, Arid Zone
    Research* **15**, 105–38. Paris. UNESCO.
Orshansky, G. (1954). Surface reduction and its significance as a
    hydroecological factor. *Journal of Ecology* **42**, 442–4.
Pearson, C. J. (1980). The effect of moisture stress on tobacco varieties.
    Ph.D. thesis. University of Nottingham.
Penman, H. L. (1948). Natural evaporation from open water, bare soil and
    grass. *Proceedings of the Royal Society, London* Series A **193**, 120–45.
Raper, C. D., & Downs, R. J. (1976). Field phenotype in phytotron
    culture: A case study for tobacco. *The Botanical Review* **42**, 317–43.
Rawlins, S. L. (1963). Resistance to water flow in the transpiration
    stream. In *Stomata and Water Relations in Plants*, ed. I. Zelitch, pp.
    69–85. Bulletin 664 of the Connecticut Agricultural Experiment
    Station, New Haven, Connecticut.
Rawson, H. H., Begg, J. E. & Woodward, R. G. (1977). The effect of
    atmospheric humidity on photosynthesis, transpiration and water use
    efficiency of leaves of several plant species. *Planta* **134**, 5–10.
Ritchie, J. T. (1972). Model for predicting evaporation from a row crop
    with incomplete cover. *Water Resources Research* **8**, 1204–13.
Ritchie, J. T. (1973). Influence of soil water status and meteorological
    conditions on evaporation from a corn crop. *Agronomy Journal* **65**,
    893–7.
Ritchie, J. T. & Burnett, E. (1971). Dryland evaporative flux in a
    subhumid climate. II. Plant influences. *Agronomy Journal* **63**, 56–62.
Sayre, J. D. (1926). Physiology of stomata of *Rumex patientia*. *Ohio
    Journal of Science* **26**, 233–66.
Schulze, E.-D. (1972). A new type of climatised gas exchange chamber
    for net photosynthesis and transpiration measurements in the field.
    Oecologia **10**, 243–51.
Schulze, E.-D., Lange, O. L., Buschbom, U., Kappen, L., & Evenari, M.
    (1972). Stomatal responses to changes in humidity in plants growing in
    the desert. *Planta* **108**, 259–70.
Schulze, E.-D., Lange, O. L., Evenari, M., Kappen, L., Buschbom, U.
    (1974). The role of air humidity and leaf temperature in controlling
    stomatal resistance of *Prunus armeniaca* L. under desert conditions.
    I. A simulation of the daily course of stomatal resistance. *Oecologia*
    17, 159–70.
Sharpe, P. J. H. (1973). Adaxial and abaxial stomatal resistances of
    cotton in the field. *Agronomy Journal* **65**, 570–4.
Shawcroft, R. W., Lemon, E. R. & Stewart, D. W. (1973). Estimation of
    internal crop water status from meteorological and plant parameters.
    In *Plant Responses to Climatic Factors,* ed. R. O. Slatyer, pp. 449–59.
    Paris: UNESCO.
Shuttleworth, W. J. (1979). *Evaporation.* Report 56, Institute of
    Hydrology, Wallingford, UK.
Sinclair, T. R., Murphy, C.E. & Knoerr, K. R. (1976). Development and
    evaluation of simplified models for simulating canopy photosynthesis
    and transpiration. *Journal of Applied Ecology* **13**, 813–29.
Specht, R. L. (1972). Water use by perennial evergreen plant
    communities in Australia and Papua New Guinea. *Australian Journal
    of Botany* **20**, 273–99.
Squire, G. R. (1979). Response of stomata of pearl millet (*Pennisetum*

typhoides S. & H.) to atmospheric humidity. *Journal of Experimental Botany* 30, 925–34.

Stern, W. R. (1965). Evapotranspiration of safflower at three densities of sowing. *Australian Journal of Agricultural Research* 16, 961–71.

Stigter, C. J. (1972). Leaf diffusive resistance to water vapour and its direct measurement. 1. Introduction and review concerning relevant factors and methods. *Mededelingen van de Landbouwhogeschool te Wageningen* 72–3, 1–47.

Stigter, C. J., Birnie, J. & Lammers, B. (1973). Leaf diffusive resistance to water vapour and its direct measurement. II. Design, calibration, and pertinent theory of an improved leaf diffusion resistance meter. *Mededelingen van de Landbouwhogeschool te Wageningen* 73–15, 1–55.

Stigter, C. J. & Lammers, B. (1974). Leaf diffusive resistance to water vapour and its direct measurement. III. Results regarding the improved diffusion porometer in growth rooms and fields of Indian corn (*Zea mays*). *Mededelingen van de Landbouwhogeschool te Wageningen* 74–21, 1–76.

Szeicz, G. & Long, I. F. (1969). Surface resistance of crop canopies. *Water Resources Research* 5, 622–33.

Szeicz, G., Van Bavel, C. H. M., & Takami, S. (1973). Stomatal factor in the water-use and dry matter production of sorghum. *Agricultural Meteorology* 12, 361–89.

Taylor, H. M. & Klepper, B. (1978). The role of rooting characteristics in the supply of water to plants. *Advances in Agronomy* 30, 99–128.

Teare, I. D., & Kanemasu, E. T. (1972). Stomatal-diffusion resistance and water potential of soybean and sorghum leaves. *New Phytologist* 71, 805–10.

Thom, A., & Oliver, H. R. (1977). On Penman's equation for estimating regional evaporation. *Quarterly Journal of the Royal Meteorological Society* 103, 345–57.

Turner, N. C. (1970). Response of adaxial and abaxial stomata to light. *New Phytologist* 69, 647–53.

Turner, N. C. (1974). Stomatal responses to light and water under field conditions. *Bulletin of the Royal Society of New Zealand* 12, 423–32.

Turner, N. C. & Heichel, G. H. (1977). Stomatal development and seasonal changes in diffusive resistance of primary and regrowth foliage of red oak (*Quercus rubra* L.) and red maple (*Acer rubrum* L.). *New Phytologist* 78, 71–81.

Turner, N. C. & Incoll, L. D. (1971). The vertical distribution of photosynthesis in crops of tobacco and sorghum. *Journal of Applied Ecology* 8, 581–91.

Van Bavel, C. H. M. (1967). Changes in canopy resistance to water loss from alfalfa induced by soil water depletion. *Agricultural Meteorology* 4, 165–76.

Van Bavel, C. H. M. (1973). Towards realistic simulation of the natural plant climate. In *Plant Response to Climatic Factors,* ed. R. O. Slatyer, *Ecology and Conservation* 5, 441–46. Paris: UNESCO.

Watts, W. R. (1977). Field studies of stomatal conductance. In *Environmental Effects on Crop Physiology,* ed. J. J. Landsberg & C. V. Cutting, pp. 173–96. London: Academic Press.

Woods, D. B. & Turner, N. C. (1971). Stomatal response to changing light by four tree species of varying shade tolerance. *New Phytologist* 70, 77–84.

P. G. JARVIS & J. I. L. MORISON

The control of transpiration and photosynthesis by the stomata

Introduction

Stomata occur on the sporophytes of both Pteridiophyta and Bryophyta and therefore are likely to have appeared early in evolution probably in association with the evolutionary transitions of early plants from an aquatic to a terrestrial habitat.

The stomata are situated in the leaf surface in the one position where they are most effectively placed to control the influx and efflux of water vapour and gases to the interior of the leaf.

Classical calculations by Gradmann (1928), revived by van den Honert (1948), demonstrate the highly effective placement of stomata in the water continuum between soil and atmosphere. Whilst the drop in water potential between soil and leaf normally lies between about 0.3 and 3 MPa (3 to 30 bar), the drop in water potential between the cells within the leaf and the atmosphere may be about 90 MPa (50% relative humidity at 20 °C). The stomata are therefore situated where the largest drop in potential in the whole soil–plant–atmosphere continuum occurs and are, therefore, in the most advantageous position to control the flow of water from soil to atmosphere. It follows that any changes in the properties of the flow pathway, leading to changes in permeability to water, will only cause a change in the rate of flow if the stomatal conductance changes as a result (Philip, 1957).

In the case of the influx of gases, such as carbon dioxide and sulphur dioxide, a substantial part of the overall drop in concentration between the atmosphere and the sites of reaction within the leaf cells may also occur across the stomatal pore. However, the pathways for gases are more complex than for water vapour since they involve intercellular transport, solubilization and enzymic reactions, such as carboxylation and decarboxylation. Consequently substantial limitation to the influx and efflux of gases also occurs within the cells (see Fig. 1). Nonetheless, the stomata are also likely to exert effective control over gaseous exchange (Unsworth, Biscoe & Black, 1976).

These are the classical views – that the stomata control effectively the fluxes of water vapour and gases between the cells within the leaf and the atmosphere.

However, stomatal aperture or conductance depends on several variables, the values of which depend upon the rates of transpiration and photosynthesis. For example stomatal conductance may depend upon the bulk leaf water content or water potential, which may itself be a function of transpiration rate (Weatherley, 1970; Jarvis, 1975). Similarly stomatal conductance in many species depends on the mean intercellular space carbon dioxide concentration (Meidner & Mansfield, 1968) which in turn depends upon the rate of photosynthesis. Thus a question arises with respect to both transpiration and photosynthesis. To what extent do the stomata control the rate of the process, or are controlled themselves by the process?

To approach this question we must first define the apparent responses of stomatal conductance to environmental variables.

Fig. 1. A model of the pathways of water vapour and carbon dioxide exchange in a photosynthesizing leaf (adapted from Lake, 1967). A is the carbon dioxide sink at the site of assimilation; B is the source of carbon dioxide in respiration. The symbols are listed in Table 1.

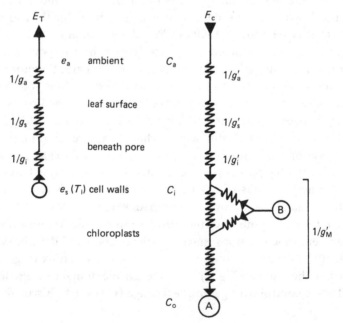

Stomatal conductance

Stomatal conductance may be defined as the proportionality parameter relating the flux of water vapour through the stomatal pore to the driving force (in consistent units)

$$g_s = \text{(flux)/(driving force)}.$$

It may also be derived from the dimensions of the pathway (see for example Cowan & Milthorpe, 1968).

There are five main environmental variables and one state variable which determine the value of g_s at a particular time: incident quantum flux density (light), temperature of the leaf, humidity of the ambient air, ambient carbon dioxide concentration and bulk leaf water potential (Burrows & Milthorpe, 1976). Fig. 2 shows idealized responses of stomatal conductance to these variables.

Fig. 2. The response of stomatal conductance to (*a*) quantum flux density (*Q*), (*b*) leaf temperature (*T*$_l$), (*c*) leaf–air vapour pressure difference or vapour pressure deficit (*D*), (*d*) mean intercellular space carbon dioxide concentration (*C*$_i$) and (*e*) cell water or xylem pressure potential in the leaves (*ψ*). Modified from Jarvis (1975).

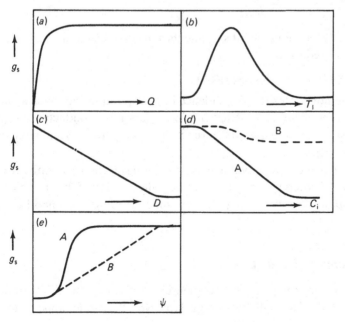

Transpiration

To what extent do the stomata control transpiration or are controlled by transpiration?

The rate of transpiration from a leaf can be written as a function of stomatal conductance in the following way

$$E_T = \frac{c_p \rho (e_s(T_1) - e_a) g_w}{\lambda \gamma}. \tag{1}$$

The symbols are listed in Table 1. g_w is the conductance for water-vapour transfer between the sites of evaporation within the leaf (see Tyree & Yianoulis, 1980) and the ambient atmosphere. The main diffusion resistance of this pathway lies within the stomatal pore so that we may write, see Fig. 1,

$$\frac{1}{g_w} = \frac{1}{g_a} + \frac{1}{g_s} + \frac{1}{g_i}$$

and $g_s \ll g_a < g_i$ (Milthorpe, 1961).

By application of the principle of the conservation of energy to the heat and water-vapour transfer equations, Penman (1948) eliminated the requirement for leaf temperature from eqn (1). Following Monteith (1965) we may write also

$$E_T = \frac{sA + c_p \rho D g_a}{\lambda(s + \gamma(1 + g_a/g_s))}. \tag{2}$$

Eqns (1) and (2) are equivalent statements and when leaf and air temperature are equal reduce to

$$E_T = kDg_s \qquad \text{(Jarvis, 1981)}.$$

Eqn (2) demonstrates the dependence of transpiration rate on available energy (A), vapour pressure deficit (D), boundary layer conductance (g_a, a function of windspeed), s (de_s/dT, a function of temperature) and stomatal conductance (g_s).

Both eqns (1) and (2) imply that g_s controls transpiration rate rather than the converse. We therefore need to consider whether any of the variables to which g_s responds, shown in Fig. 2, determine, or are determined by, the transpiration rate.

Quantum flux density

Stomata open in response to quantum flux density and transpiration increases in relation to available energy. However, the former appears to be

Table 1. *A list of the symbols for quantities used and their units*

A	available energy	J m^{-2} s^{-1}
c_p	specific heat of air at constant pressure	J kg^{-1} °C^{-1}
C_a	ambient carbon dioxide concentration	cm^3 m^{-3}
C_i	mean carbon dioxide concentration in the intercellular spaces	cm^3 m^{-3}
C_o	carbon dioxide concentration at the site of carboxylation in the mesophyll	cm^3 m^{-3}
D	leaf–air vapour pressure difference, vapour pressure deficit	kPa
E_T	total transpiration rate $(=E_M+E_C)$	mg m^{-2} s^{-1}
E_M	rate of transpiration from sub-stomatal cavity	mg m^{-2} s^{-1}
E_C	rate of extra-stomatal transpiration	mg m^{-2} s^{-1}
e_a	ambient vapour pressure	kPa
$e_s\,(T_1)$	saturation vapour pressure at leaf temperature	kPa
F_c	rate of net photosynthesis	mg m^{-2} s^{-1}
$F_{c,max}$	light- and carbon dioxide-saturated rate of net photosynthesis	mg m^{-2} s^{-1}
g_W	leaf conductance for water vapour transfer	cm s^{-1}
g_C'	leaf conductance for carbon dioxide transfer	cm s^{-1}
g_a, g_a'	boundary layer conductance for water vapour and carbon dioxide, respectively	cm s^{-1}
g_s, g_s'	stomatal conductance for water vapour	cm s^{-1}
g_i, g_i'	intercellular space conductance for water vapour and carbon dioxide, respectively	cm s^{-1}
G	stomatal conductance normalized with respect to the maximum value in the experiment	dimensionless
g_M'	mesophyll (or residual) conductance for carbon dioxide	cm s^{-1}
g_g'	gas phase conductance for carbon dioxide	cm s^{-1}
k	coefficient for converting water vapour pressure to absolute humidity $(=\rho c_p/(\gamma\lambda))$	mg m^{-3} kPa^{-1}
Q	quantum flux density	μE m^{-2} s^{-1}
R_E, R_M, R_H	flow resistances for liquid water in the endodermis, mesophyll and hypodermis, respectively	Pa s m^{-3}
s	rate of change of saturated water vapour pressure with respect to temperature (de_s/dT)	kPa °C^{-1}
ρ	density of air	mg m^{-3}
ρ_c	density of carbon dioxide	mg m^{-3}
γ	psychrometric constant	kPa °C^{-1}
Γ	carbon dioxide compensation concentration	cm^3 m^{-3}
ψ	leaf water potential, xylem pressure potential	MPa

a quantum response related to photochemical processes, and the latter is clearly a response to the availability of energy. It seems intrinsically unlikely that the two responses are mechanistically related. However, they will inevitably be correlated since the incident quantum flux density and, to a lesser extent, the flux density of available energy, are both closely dependent on the flux density of solar radiation.

Ambient carbon dioxide

The process of transpiration is not affected by carbon dioxide other than through its effect on g_s.

Temperature

Neither g_s nor transpiration rate are very sensitive to air temperature. While transpiration tends to increase with temperature, g_s usually has an optimum temperature, as shown in Fig. 2 (Burrows & Milthorpe, 1976). It is likely that this optimum temperature for g_s is related to the temperature characteristics of certain of the enzymes involved in stomatal action (Rogers, Powell & Sharpe, 1979). In addition vapour pressure deficit usually increases with a rise in temperature and this will also result in stomatal closure at supra-optimal temperatures (see below). The contrasting tendency for transpiration rate to increase with temperature is partly the result of the effect of temperature on the rate of change of saturation vapour pressure with respect to temperature (i.e. s) and partly the result of the tendency for vapour pressure deficit to increase at higher temperatures. However, if g_s falls substantially at higher temperatures, transpiration rate may also show an optimum response to temperature (Landsberg & Butler, 1980).

Ambient humidity

Transpiration rate tends to increase with increase in the vapour pressure deficit, whereas stomatal conductance in many species tends to decrease. This can be the result of negative feedback between transpiration rate and g_s consequent upon a fall in leaf water potential but other observations suggest that this is frequently not the case. For example, g_s continues to fall at large vapour pressure deficits even though the transpiration rate from the leaf as a whole is also falling (Fig. 3). In addition, the decline in g_s with increasing vapour pressure deficit may occur whilst the bulk leaf water potential remains constant (e.g. Watts & Neilson, 1978). These observations suggest that g_s responds to humidity independently of the effect of humidity

Fig. 3. The response of (*a*) relative stomatal conductance and (*b*) transpiration rate to increasing leaf–air vapour pressure difference (*D*) in Scots pine (*Pinus sylvestris* L.) at temperatures of 10 °C (triangles), 20 °C (circles) and 25 °C (squares). Average g_s at $G = 1$ was 0.43 cm s^{-1}; quantum flux density was 1320 μE m^{-2} s^{-1}. The measurements have been normalized to reduce variation between shoots. All points are means of 5 measurements with $\pm$ 1 s.e. shown on representative points. From Ng (1978).

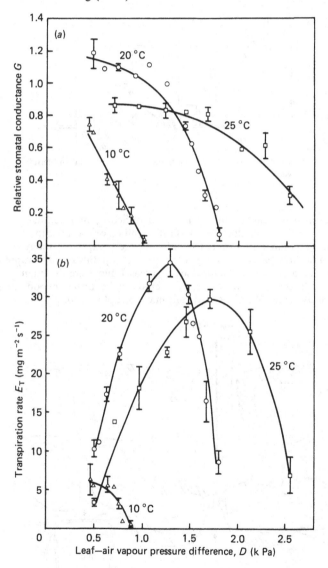

Fig. 4. A diagram of the liquid (single lines) and vapour flow (double lines) pathways of water in a leaf to illustrate the hypothesis for the response to vapour pressure deficit. Liquid flow resistances R_E, R_M and R_H in the pathway between the xylem and stomata cause a drop in potential so that the stomatal apparatus is sensitive to small changes in the extrastomatal vapour flux (E_c) which result from changes in the vapour pressure deficit. s.c. subsidiary cell; g.c. guard cell; symbols in Table 1. Adapted from Ng (1978).

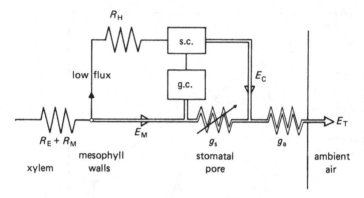

Fig. 5. A transverse section through a stoma of Scots pine (modified from Ng, 1978). The shaded areas indicate lignified cell walls with unshaded, unlignified areas in between from which evaporation may occur. Liquid flow pathways, full lines; vapour flux pathways, interrupted lines. Areas 2 and 3 are the 'hinge' areas; areas 1 and 3 are the main sources of the evaporation flux, E_C; e.c. epidermal cell; h.c. hypodermal cell; s.c. subsidiary cell; g.c. guard cell; m.c. mesophyll cell.

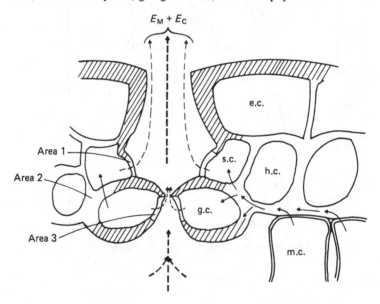

on the rate of transpiration from the leaf as a whole (Sheriff & Kaye, 1977; Lösch, 1977; Jarvis, 1980).

The results in Fig. 3 suggest that g_s is responding directly to vapour pressure deficit as a result of a water vapour flux which occurs external to the stomatal pore (Fig. 4) (Farquhar, 1978). For stomatal turgor to be sensitive to such a water vapour flux, there needs to be an appreciable resistance to liquid flow to the stomata (Sheriff, 1977), as shown in Fig. 4. A possible site for this resistance in Scots pine is in the lignified walls of the hypodermis: a possible site for the source of the extra stomatal vapour flux in Scots pine is the anticlinal wall between the guard cells and subsidiary cells (Fig. 5) (Jarvis, 1980).

Such a response of g_s to large vapour pressure deficits reduces transpiration rate and prevents the development of low water potentials in the leaf thereby avoiding stress (e.g. Schulze *et al.*, 1972, 1974). However, the reduction in g_s does not occur as the result of an effect of transpiration from the leaf as a whole on g_s.

Leaf water potential

Transpiration from leaves leads to the development in the leaves of the driving force for the movement of liquid water through the plant – the reduction in leaf water potential (ψ). This seems to be the most likely effector of feedback between E_T and g_s. Two kinds of response of ψ to E_T have been observed (Fig. 6; see review by Jarvis, 1975):

Fig. 6. A diagram to show the alternative relations between leaf water potential (ψ) and transpiration rate (E_T) which have been found. A, the decrease in ψ is proportional to the increase in E_T; B, an initial fall in ψ is followed by a constant value which implies a constant potential difference across the plant and hence a variable resistance. Modified from Jarvis (1975).

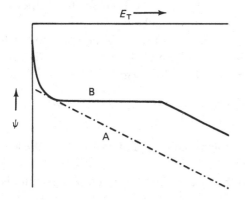

A) as E_T increases, ψ declines,

B) as E_T increases, ψ falls a little but then remains constant.

The first type of response (A) seems to be restricted to woody plants, whereas response (B) has been observed in a wide range of species (Jarvis, 1975).

In the case of (B) no feedback through ψ is possible and therefore the response of g_S to ψ is irrelevant. In the case of (A) feedback is possible but the degree of control which occurs depends on the type of relation between g_s and ψ. Fig. 2e shows two common types of relationship. In the case of the 'threshold' response (A), which is common in field crops, no significant reduction in g_s occurs until ψ reaches a very low level: in the case of the 'linear' response (B), stomatal closure will occur as a result of the normal daily fluctuations in ψ (see Burrows & Milthorpe, 1976; Beadle, Jarvis & Neilson, 1979; Turner, 1979).

Thus, if ψ declines with increasing E_T and if g_s declines with decreasing ψ, g_s will change in response to changes in transpiration rate. This combination of circumstances is not encountered as often as was previously thought, especially in field-grown plants.

Conclusion

Stomatal conductance is largely independent of transpiration rate from the leaf as a whole. At times g_s may seem to increase with E_T because both depend on solar radiation but the parallelism is not the result of a common mechanism. At times g_s may seem to decrease as E_T increases because of an inverse response of g_s to vapour pressure deficit. Whilst both E_T and g_s respond to vapour pressure deficit through changes in water vapour fluxes, these fluxes have different consequences, probably because they originate at separate sites. Consequently the association between the response of E_T and g_s to vapour pressure deficit is correlative rather than based on a process common to both. In many plants g_s only comes to depend on E_T in stress conditions when high rates of transpiration lead to low leaf water potentials which feed back to reduce g_s. Thus in general E_T can be regarded as controlled by g_s rather than *vice versa*.

Photosynthesis

To what extent do the stomata control photosynthesis by the leaf or are controlled by photosynthesis?

In 1959 Gaastra expressed the rate of light saturated photosynthesis as a function of the resistances in the carbon dioxide transfer pathway. In an analogous form to eqn (1), the rate of photosynthesis was expressed as

$$F_c = \rho_c(C_a - C_o)g_C',$$

$$= \rho_c(C_a - C_i)g_g' = \rho_c(C_i - C_o)g_M',$$

where $\dfrac{1}{g_g'} = \dfrac{1}{g_a'} + \dfrac{1}{g_s'} + \dfrac{1}{g_i'}$

and, as before, $g_s' \ll g_a' < g_i'$.

The symbols are listed in Table 1. g_C is the conductance for carbon dioxide transfer between the ambient air and the sites of carboxylation in the chloroplast. The main diffusion resistance in this pathway in the gas phase lies within the stomatal pore but there is also a substantial liquid phase transfer resistance between the cell walls and the chloroplasts of the mesophyll cells so that when the stomata are open $g_M' < g_g'$ (Jarvis, 1971).

By taking into account the hyperbolic form of the response of photosynthesis to both quantum flux density and carbon dioxide concentration, the following expression for F_c may be obtained (Thornley, 1976; Reed, Hamerly, Dinger & Jarvis, 1976)

$$F_c = \frac{Qg_M'\rho_c(C_i - \Gamma)}{\alpha Q + g_M'\rho_c(C_i - \Gamma) + \alpha Qg_M'\rho_c(C_i - \Gamma)/F_{c,max}} \tag{4a}$$

where $C_i = C_a - F_c/(\rho_c g_g')$. $\tag{4b}$

Eqns (4 *a, b*) demonstrate the dependence of F_c on quantum flux density (Q), ambient carbon dioxide concentration (C_a), mesophyll conductance (g_M') and the gas phase conductance which is predominantly stomatal (g_s').

Both eqns (3) and (4*a, b*) imply that g_s' controls the rate of photosynthesis rather than the converse. We must therefore consider whether any of the variables shown in Fig. 2, to which g_s' responds, determine, or are determined by the photosynthetic rate.

Ambient humidity

The response of g_s (or g_s') to vapour pressure deficit is probably independent of photosynthesis. As far as is known, photosynthesis is not affected by vapour pressure deficit, other than through its effects on g_s.

Leaf water potential

The response of g_s to leaf water potential (ψ) (Fig. 7) is probably also largely independent of photosynthesis. Whilst photosynthesis may be reduced at moderately low leaf water potentials, in most cases this effect can largely be attributed to stomatal closure rather than to inhibition of photosynthetic processes by the stress (Hsiao, 1973). However, the photochemical and enzymatic partial processes of photosynthesis are inhibited by low water potentials (Keck & Boyer, 1974; Boyer, 1976) and this may lead to reduction in g_s. Because the reduction in photosynthesis at low water potentials can be attributed to both inhibition and stomatal closure, it is difficult to determine whether the decrease in g_s is causally related to photosynthesis or *vice versa*.

Temperature

In temperate plants both g_s and F_c respond to moderate temperatures in a parallel way and show an optimum response curve. This correla-

Fig. 7. The relation between stomatal conductance (g_s) and xylem water potential ψ in forest shoots of Sitka spruce (*Picea sitchensis* (Bong.) Carr.) at five different carbon dioxide concentrations. The curves are fitted exponentials to data from 8 shoots at each C_a. Quantum flux density 1000 μE m^{-2} s^{-1}, leaf temperature 20 °C; leaf-air vapour pressure difference 0.6 kPa. From Beadle *et al* (1979).

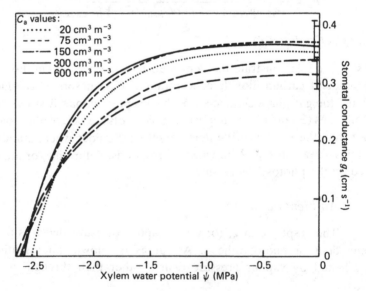

tion can be interpreted to suggest either control of the stomata by photosynthesis or the converse. Midday closure has been attributed to a rise in C_i resulting from the effect of high temperatures on respiration (Heath & Orchard, 1957) and can be reversed by flushing the intercellular spaces with carbon dioxide-free air (Meidner & Heath, 1959). Such observations have led to the conclusion that the response of g_s to temperature is the result of changing rates of *net* photosynthesis and hence of C_i. However g_s may not correlate well with F_c or C_i over a wide range of temperatures. In Sitka spruce, for example, opposite relationships between g_s and C_i are found above and below the optimum temperature for g_s (Neilson & Jarvis, 1975) and in desert and temperate plants exposed to high temperatures the stomata often open widely at times when F_c is falling and C_i rising. Moreover, g_s responds to temperature in both light and darkness in carbon dioxide-free air (Rogers, Powell & Sharpe, 1979; Rogers, Sharpe & Powell, 1980) and in darkness in normal air (Mansfield, 1965; Brunner & Eller, 1974). Thus the effects of temperature on g_s cannot simply be mediated through changes in F_c and C_i. It seems probable that both F_c and g_s are responding through independent processes.

Quantum flux density

Both g_s and F_c increase with quantum flux density, often with a similarly shaped hyperbolic response. The resulting correlations can be interpreted to suggest that g_s depends on the rate of photosynthesis by the leaf and there have been several explanations of stomatal action put forward along these lines. For example, photosynthesis by the mesophyll reduces C_i and stomatal aperture can be manipulated by flushing the intercellular spaces with air containing different carbon dioxide concentrations. As a result, an early suggestion was that photosynthesis by the mesophyll controls g_s by effecting changes in C_i (Heath, 1948). Another suggestion is the production in the mesophyll by photosynthesis of intermediates which are necessary for stomatal action, and their translocation to the guard cells. These intermediates might be either carbon products of photosynthesis or photorespiration (e.g. glucose and sucrose (Dittrich & Raschke, 1977) or glycine and serine (Willmer, Thorpe, Rutter & Milthorpe, 1978)).

Alternatively both stomatal action and leaf photosynthesis could depend on light through wholly independent mechanisms, as was supposed in the case of g_s and E_T. Support for this suggestion comes from observations that stomata will open and close in response to light in some species in the presence or absence of carbon dioxide (Fig. 8). This suggests that stomatal

action is not coupled to photosynthetic carbon fixation, since that is negative
in carbon dioxide-free air, or indeed to C_i, since similar values of g_s result
at quite different values of C_i in such experiments. Dependence of
g_s on products of the photosynthetic light reactions translocated
from the mesophyll to the guard cells is not excluded by this argument,
but there is no evidence to suggest that this is likely and opening and
closing in isolated epidermis in response to light suggests that this is not
necessary.

Another argument which has been used to support the dependence of g_s
on mesophyll photosynthesis is a general similarity between the action
spectra for stomatal opening and photosynthesis. This has been taken to
imply a close relationship between the two processes. However, stomatal
opening seems to be much more sensitive to the blue end of the spectrum
than is consistent with a system driven solely by photosynthesis (Kuiper,
1964; Hsiao, Allaway & Evans, 1973). Such observations have stimulated a
number of experiments on the effects of blue light on stomatal action. In
Scots pine, for example, at an incident quantum flux density of 100 μE m^{-2}
s^{-1}, F_c in red light was three times that in blue light, whereas g_s in blue light

Fig. 8. The relation between relative stomatal conductance (G)
(normalized relative to the maximum value for each shoot) and
increasing quantum flux density in forest shoots of Scots pine in air of
the following carbon dioxide concentrations $C_a \simeq 340$ cm^3 m^{-3}
(circles), $= 30$ cm^3 m^{-3} (squares), $\simeq 5$ cm^3 m^{-3} (triangles), two or three
replicates per treatment; ± 1 S.E. shown on representative points.
Average g_s at $G = 1$ was 0.52 cm s^{-1}. Leaf temperature 10 °C; leaf–air
vapour pressure difference 0.55 kPa. From Ng (1978).

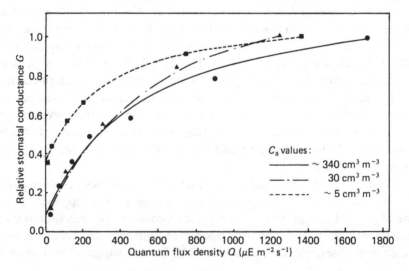

was three times that in red light although F_c was still negative (Fig. 9). Such observations are not consistent with a close mechanistic dependence of g_s on mesophyll photosynthesis.

Carbon dioxide

Since C_i declines as F_c increases and since g_s in many species increases with decreasing C_i, it has been supposed that mesophyll photosynthesis controls g_s by effecting changes in C_i. However, whether photosynthesis by the mesophyll is directly involved is questionable.

There are two sets of observations which suggest more strongly than anything else that stomatal action depends upon the concentration of carbon dioxide in the vicinity of the guard cells rather than on mesophyll photosynthesis itself. In the first place, stomatal closure as a result of the inhibition of photosynthesis by metabolic inhibitors can be reversed by forcing carbon dioxide-free air through the leaf (Allaway & Mansfield, 1967). Secondly, stomatal opening in the dark can be induced in the same way and closure can be induced by forcing carbon dioxide at high concentration through the leaf in the light (Mouravieff, 1965; Mansfield, 1965). These experimental results strongly suggest that in some situations, at least, there is

Fig. 9. The relation between (a) relative stomatal conductance (G) and (b) rate of net photosynthesis (F_c) and quantum flux density of broad band red (peak 660 nm) (open circles) or blue light (peak 420 nm) (filled circles) in Scots pine. Four replicate shoots; individual points shown. Average g_s at $G = 1$ was 0.28 cm s^{-1}. Leaf temperature 20 °C; leaf–air vapour pressure difference 0.7 kPa. From Morison (1980).

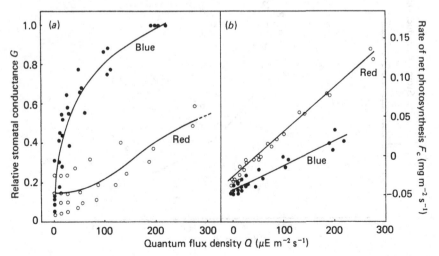

a carbon dioxide-sensitive system which has a controlling influence on stomatal action. Although mesophyll photosynthesis itself is only indirectly involved, some feedback control of g_s by mesophyll photosynthesis may result (Farquhar, Dubbe & Raschke, 1978).

In many experiments in which F_c and g_s have been measured, particularly those in which quantum flux density was the variable, it has been found that C_i reaches a fairly constant value (between 200 and 220 cm³ m⁻³) at high quantum flux densities (Fig. 10). It has been argued from such observations

Fig. 10. The relation between stomatal conductance (g_s') and mean intercellular space carbon dioxide concentration (C_i) as quantum flux density was raised from darkness to 700 μE m⁻² s⁻¹ in three types of leaves of Golden Delicious apples. Leaf temperature 21 °C; leaf–air vapour pressure difference 1.2 kPa. From Warrit (1977).

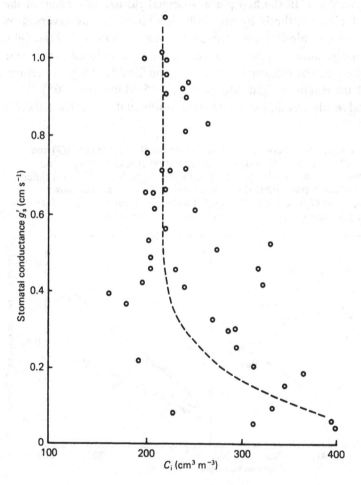

that a C_i of about 200 cm³ m⁻³ is the optimum setpoint of a stomatal control system which is sensitive to carbon dioxide, so that if photosynthesis increases, giving a tendency for C_i to decrease further, the stomata open further to maintain the same C_i (Farquhar *et al.*, 1978). Cowan and Farquhar (1977) suggested that evolution might have resulted in the optimization of the properties of such a control system to ensure that the maximum amount of carbon dioxide was fixed per unit of water transpired, and that a more or less constant value of C_i was a consequence of this.

From eqns (3) and (4*b*) it can readily be shown in the absence of photo-respiration that

$$C_i = C_a \frac{g_g'}{g_g'+g_M'}. \tag{5}$$

That is to say, for a particular C_a, C_i will be constant provided that g_g' and g_M' change together, in response to light for example, so that g_g'/g_M' remains constant. This is intuitively obvious from the resistance analogue in Fig. 1: the potential in the middle of the resistance catena will remain constant if the resistances remain in fixed proportion to one another and the drop in potential over the whole does not change. Thus a constant C_i could be taken as evidence of interdependence between g_s' (the major component of g_g') and g_M'. Such interdependence, it has been postulated, would take the form of a mechanistic coupling or negative feedback between g_M' and g_s', so

Table 2. *The rate of change of C_i with respect to C_a (i.e. $\Delta C_i/\Delta C_a$) over the range of C_a, 50 – 500 cm³ m³, for a number of species with both carbon dioxide-sensitive (+) and carbon dioxide-insensitive (−) stomata. The measurements were made under controlled conditions in assimilation chambers. The following sources give the original data: (1) Dubbe, Farquhar & Raschke (1978), (2) Farquhar, Dubbe & Raschke (1978), (3) Warrit, Landsberg & Thorpe (1980), (4) Wong, Cowan & Farquhar (1979), (5) Morison (1980).*

Species	$\Delta C_i/\Delta C_a$	CO_2-sensitivity	Source
Avena sativa L.		+	1
Commelina communis L.	0.88–0.92	+	5
Gossypium hirsutum L.		+	1
Xanthium strumarium L.		+	1,2
Eucalyptus pauciflora Sieb. ex Spreng		+	4
Malus pumila Mill. (cv)	0.76–0.80	+	3
Picea sitchensis (Bong.) Carr.		−	5
Pinus sylvestris L.		−	5
Amaranthus powelli S. Wats	0.7	+	1
Zea mays L.	0.4	+	2,4

that if g'_M and hence F_c change in response to changes in light or temperature, for example, proportional changes will occur in g'_s and C_i will not
change. Apparently constant C'_i as shown in Fig. 10, could therefore be
regarded as evidence of feedback control of g_s by F_c (Farquhar *et al.*, 1978).

However, such a constant C_i is only obtained in steady-state conditions
and not in the fluctuating conditions which occur in the field. The response
time of the instantaneous rate of photosynthesis to changes in quantum flux
density may be six orders of magnitude faster than the response time of the
stomata. The response of C_i is somewhat slower because of the reservoir of
carbon dioxide in the leaf, but nonetheless, C_i fluctuates widely in the field
while g_s remains relatively constant. Consequently in field conditions there is
no consistent relationship between g_s and C_i (Fig. 11) (see also Schulze *et al.*,
1975). In these circumstances it is difficult to maintain a case for a unique
role for particular values of C_i.

Secondly, such an apparently constant C_i is not maintained when C_a is
varied outside the normal range, or if F_c is caused to vary by other means such
as the use of metabolic inhibitors (e.g. Wong, Cowan & Farquhar, 1979).

Fig. 11. Stomatal conductance of current year needles of Douglas fir
(*Pseudotsuga menziesii* (Mirb.) Franco) in a forest canopy, plotted in
relation to C_i. The measurements were made on five similar shoots at the
ends of branches using assimilation chambers over several days of
varying weather at Cedar River in Washington State. See Leverenz
(1980), for details.

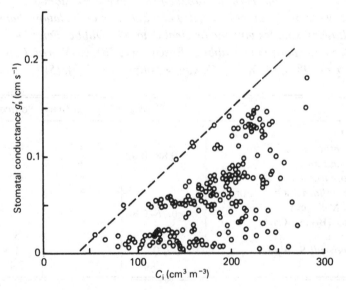

From eqn (5) it follows that C_i/C_a will be constant if g_s'/g_M' is constant. C_i will only remain constant as C_a is changed if proportional changes occur in g_s'/g_M': for a decrease in C_a, g_s'/g_M' must increase, and *vice versa*, if C_i is to remain constant. In practice, the relation between C_i and C_a remains reasonably constant over a considerable range of C_a with a slope which expresses the relative magnitudes of the gas and liquid phase conductances (Table 2). These data show that the gain in the negative feedback loop between F_c and g_s is insufficient to keep C_i constant ($\delta C_i/\delta C_a \neq 0$) and hence that mesophyll photosynthesis influences g_s' through C_i to only a small extent (Farquhar *et al.*, 1978). Wong *et al.* (1979), therefore, invoked translocation of metabolites from mesophyll to guard cells to explain their observed correlations between F_c and g_s', making the tacit assumption that the correlations implied a mechanistic relationship. It has already been pointed out, however, that the correlation between g_s and F_c can readily be broken, in carbon dioxide-free air for example.

Furthermore, in other species or conditions, g_s is insensitive or shows very low sensitivity to C_i in the normal physiological range. Whilst the stomata of many species are undoubtedly sensitive to carbon dioxide, the stomata of others are not (e.g. Parkinson 1968; Neilson & Jarvis, 1975; Beadle *et al.*, 1979) and, in addition, species with normally carbon dioxide-insensitive stomata, may at times have carbon dioxide-sensitive stomata (e.g. Pallas, 1965). Raschke (1975) provided evidence that a response to carbon dioxide in *Xanthium strumarium* L. was only elicited after the leaves had been water stressed so that abscisic acid had accumulated. However, Mansfield (1976) found little or no interdependence between the action of abscisic acid and carbon dioxide and Beadle *et al.* (1979) did not find a sensitivity to carbon dioxide induced as a result of water stress in Sitka spruce (Fig. 7). In other cases, too, variations in sensitivity of g_s to C_i were not related to water stress (Warrit, 1977; Goudriaan & van Laar, 1978). Clearly in the absence of sensitivity of g_s to carbon dioxide, mesophyll photosynthesis cannot be controlling g_s through changing C_i.

Conclusion

These arguments lead to the conclusion that stomatal conductance is largely independent of the rate of carbon fixation in the leaf as a whole. Both g_s and F_c often change in response to changes in temperature and light in a parallel manner. However, correlations between g_s and F_c break down in light and darkness at low carbon dioxide concentrations and in red and blue light. The insensitivity of g_s to C_i in some species and conditions, the opening of stomata in response to light in the presence or absence of carbon dioxide, and the wide range in values of C_i which occur in the field or can be induced by changing C_a experimentally suggest that there are quite separate effects of light and carbon dioxide on the guard cells and that these effects are largely unrelated to mesophyll photosynthesis. Although correlations between g_s and F_c are readily obtained, the evidence for a functional dependence of g_s on mesophyll photosynthesis is unconvincing, whereas there can be no doubt that F_c must depend on g_s.

Separate effects of light and carbon dioxide

Stomata open and close in response to external stimuli in isolated epidermal strips (e.g. Fischer, 1968; Lösch, 1977, 1979; Travis & Mansfield, 1977, 1979). Consequently possible separate effects of light and carbon dioxide on g_s must be considered in relation to the processes taking place in the guard cells and epidermal cells rather than in the leaf as a whole, as has been considered up to this point.

Although photosynthesis by the mesophyll may be largely irrelevant to stomatal action, photochemical and carbon fixation processes in the guard cells themselves are likely to be of considerable importance: the products of photochemical processes are likely to be required by the ion pumps and products of carbon fixation, such as malate, are needed to maintain charge balance (Hsiao, 1976; Raschke, 1979). Thus the physiological evidence for separate effects of light and carbon dioxide has to be considered in relation to guard cell photosynthesis rather than to mesophyll photosynthesis. Perhaps the most convincing piece of evidence for separate effects of light and carbon dioxide on processes in the guard cell is that similar values of g_s can be obtained over a wide range of values of C_i. Some examples of this will now be discussed.

Response of g_s to light with and without carbon dioxide

The opening of stomata in response to quantum flux density in the presence or absence of carbon dioxide, shown in Fig. 8, not only demonstrates the lack of dependence of g_s on F_c but is also evidence of a distinct lack of dependence of g_s on C_i. In air containing normal carbon dioxide g_s of 0.45 cm s^{-1} was obtained at a C_i of 210 cm^2 m^{-3} whereas in carbon dioxide-free air the same value of g_s was obtained at a C_i of 15 cm^3 m^{-3}. A similar lack of correlation between g_s and C_i is also found during closure in air as compared with carbon dioxide-free air. In addition hysteresis in the opening and closing responses of g_s to quantum flux density (Ng & Jarvis, 1980) leads to identical values of g_s at quite different values of C_i in air with the same carbon dioxide content. For example, during opening in normal air (Fig. 8) a g_s of 0.3 cm s^{-1} occurred at a C_i of 233 cm^3 m^{-3} whereas during closure the same value of g_s occurred at a C_i of 325 cm^3 m^{-3}.

Light saturation of g_s

Light saturation of g_s and F_c can occur at quite different quantum flux densities. For example, in 'sun' and 'shade' shoots of Sitka spruce, light saturation of g_s occurred at *ca* 300 μE m^{-2} s^{-1} whereas F_c was not light saturated at 1200 μE m^{-2} s^{-1} (Leverenz & Jarvis, 1979). As a result, g_s' of 'sun' shoots was 0.40$\pm$0.01 cm s^{-1} and g_s' of 'shade' shoots was 0.28$\pm$0.02 cm s^{-1} while C_i varied between 200 and 310 cm^3 m^{-3} (i.e. $dg_s'/dC_i = 0$ for 200$<C_i<$310). These results contrast markedly with the apparently constant C_i in Fig. 10 and again suggest that g_s does not depend on particular values of C_i.

Response of g_s to red and blue light

Earlier work with broad spectral bands of light led to the suggestion that there are two separate light effects on stomata: a red light photosynthetic effect and a blue light effect not related to photosynthesis (e.g.

Fig. 12. The relation between relative stomatal conductance and mean intercellular space carbon dioxide concentration (C_i) in Scots pine at a quantum flux density of 230 μE m^{-2} s^{-1} in broad band red or blue light. Five shoots were exposed to seven values of C_a in red and blue light. Two hours were allowed for equilibration at each C_a. Average stomatal conductance at $G = 1$ was 0.53 cm s^{-1} Leaf temperature 20 °C and leaf–air vapour pressure difference 0.7 kPa. From Morison (1980).

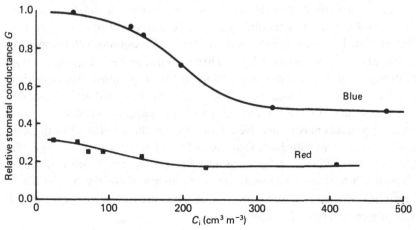

Mansfield & Meidner, 1966). As in white light, we have found that stomata will open and close in response to red or blue light in the presence or absence of carbon dioxide in both carbon dioxide-insensitive species, such as Scots pine and Sitka spruce, and in the carbon dioxide-sensitive species *Commelina communis*.

Fig. 9 shows that in Scots pine g_s is much higher at a particular quantum flux density in blue light than in red light. Since the rate of photosynthesis is much lower in blue light than in red light at the same quantum flux density, g_s was much higher in blue light than in red light at the same value of C_i (Fig. 12). [In this carbon dioxide-insensitive species somewhat higher g_s was obtained at low carbon dioxide concentrations than at high concentrations, but only after long (>2 h) periods of exposure.] These results suggest that not only are mesophyll photosynthesis and C_i of secondary importance, but blue and red light are probably acting on different and possibly separate systems in the guard cells. Since carbon metabolism seems not to be directly involved, at least in the absence of carbon dioxide, the effects of both red and blue light are probably the result of photochemical electron transport. The effect of red light may well result from photosynthetic electron transport as previously postulated (e.g. Hsiao *et al.*, 1973). The substantially higher effectiveness of blue light than red light is consistent with the hypothesis of a blue-light absorbing pigment bound to the cell membranes and driving an electron-transport chain in the cell membranes at the sites of the ion fluxes (Zeiger & Hepler, 1977).

Response of g_s to C_i at constant light

By varying C_a the response of g_s to C_i at a number of constant quantum flux densities can be determined. Fig. 13 shows such response surfaces for a carbon dioxide-insensitive and a carbon dioxide-sensitive species. In the case of Scots pine (*a*) there is little effect of C_i on g_s at any quantum flux density: g_s is independent of C_i. In *C. communis* (*b*), however, g_s responds to C_i at constant light. Furthermore, at higher quantum flux densities g_s is higher at the same C_i than at lower quantum flux densities. Thus these results also suggest independent action of light and C_i and the lack of an effect of C_i in the case of Scots pine suggests strongly that two separate processes may be involved. Nonetheless, these results, like those of a similar experiment with *Eucalyptus pauciflora* Sieb. ex Spreng (Wong *et al.*, 1979), show an interaction between the effects of light and carbon dioxide: the sensitivity to C_i increases with increasing quantum flux density. Travis & Mansfield (1979) also came to the same conclusion as a result of experiments with epidermal strips of *C. communis*.

Response of g_s to light at constant C_i

Experiments in which C_a is kept constant and quantum flux density is varied suffer from the disadvantage that C_i, varies in relation to the rate of photosynthesis and is not therefore constant. Consequently both variables of interest, light and C_i vary at the same time: a more conclusive

Fig. 13. The relation between relative stomatal conductance and mean intercellular space carbon dioxide concentration (C_i) in (*a*) Scots pine, a carbon dioxide-insensitive species and (*b*) *Commelina communis* L., a carbon dioxide-sensitive species. The numbers beside the curves are the quantum flux densities in $\mu E\ m^{-2}\ s^{-1}$. Leaf temperature 21 °C; leaf–air vapour pressure difference 0.6 kPa. Average g_s at $G = 1$ was 0.33 and 2.40 cm s^{-1} in (*a*) and (*b*) respectively. (*a*) from Ng (1978); (*b*) from Morison (1980).

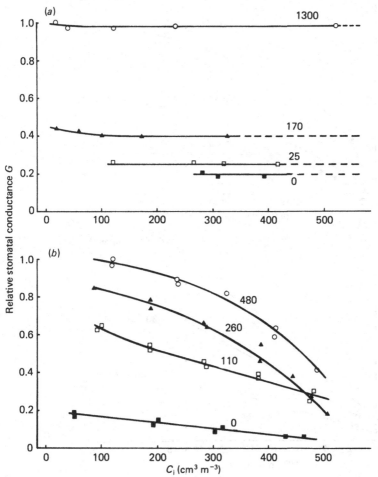

experiment is to determine the response of g_s to light at constant C_i. Heath & Russell (1954) and Lake & Slatyer (1970) tried to maintain C_i constant by flushing the intercellular spaces of leaves with air containing specified carbon dioxide concentrations. However, it is difficult to push air through the leaf fast enough to prevent the development of substantial differences

Fig. 14. (*a*) The relation between stomatal conductance (g_s) and incident quantum flux density (Q) in *Commelina communis* at three levels of 'constant' C_i ($\pm$ 1 s.e.). Leaf temperature 20 °C; leaf–air vapour pressure difference 0.26 kPa. (*b*) A double reciprocal plot of the same data. From Morison (1980).

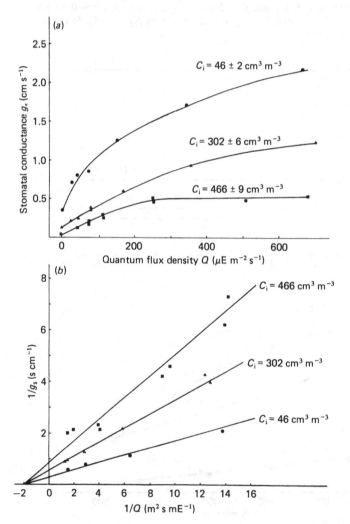

in carbon dioxide concentration across the leaf. An alternative procedure is to calculate C_i every few minutes and to adjust C_a to keep C_i within narrow limits.

Fig. 14a shows the results of a set of experiments in which the stomata were opened in response to light at three different values of C_i in a carbon dioxide-sensitive species. C_i was effectively constant over each light response curve. Similar 'constant C_i' light response curves for *Zea mays* L., another carbon dioxide-sensitive species, are shown by Raschke (1979, Fig. 18). In *C. communis* g_s clearly responded to quantum flux density at constant C_i, and with greater sensitivity at low values of C_i. Thus these results further support the hypothesis of separate effects of light and carbon dioxide on g_s.

Like Fig. 13, the results also indicate an interaction between the effects of light and C_i and this conclusion was also arrived at by Heath & Russell (1954) from the experiments in which they flushed the intercellular spaces of leaves of wheat with air of constant carbon dioxide content. The nature of this interaction can be analysed by analogy with enzyme kinetics regarding light as a substrate for stomatal action and carbon dioxide as an inhibitor. A Lineweaver-Burke plot of the same data (Fig. 14b) indicates non-competitive inhibition by carbon dioxide (Morris 1968) and thus reinforces the hypothesis of independent sites of action of light and carbon dioxide. The similar data presented by Wong, Cowan & Farquhar (1978) for *Eucalyptus pauciflora* when plotted in this way also indicate non-competitive inhibition over a limited range of C_i from *ca* 80 to 250 cm^3 m^{-3}. More determinations of the response surface of g_s to both light and C_i in carbon dioxide-sensitive species are needed to enable separation of the effects of light and C_i and the analysis of the interaction between them. The response surface for g_s in CO_2-insensitive species is unhelpful in this respect as a similar analysis cannot be made. However, an insignificant response, such as that of Scots pine in Fig. 13a, does serve to emphasise the conclusion from the above analysis that light and carbon dioxide act independently and probably at separate locations.

Conclusions

These results show that values of g_s are not correlated with particular values of C_i in a number of circumstances. In particular, the last experiment (Fig. 14), strongly supports the suggestion that light and carbon dioxide have separate, independent effects on stomatal action. In addition the experiments using red and blue light support the suggestion of two photoreceptors each contributing independently to stomatal action.

Thus in unstressed conditions, stomatal action is controlled by photo-synthetic processes and possibly also by a blue light photoreceptor. There seem to be at least two distinct components to the dependence of stomatal action on photosynthesis, the relative importance of which varies consider-ably between species and also within species depending on the conditions and previous history. In some species and conditions, stomatal action is to some extent dependent on the carbon dioxide concentration around the guard cells and this is influenced by carbon fixation in the mesophyll. However, stomatal conductance is independent of the absolute value of C_i and in other species or conditions may be almost altogether independent of C_i. The increase in g_s in response to light in such circumstances is probably the direct result of photosynthetic electron transport and photosynthetic and oxidative phosphorylation in the guard cells and their effects on ion fluxes. The balance between the control of stomatal action by carboxylation in the mesophyll or by oxidative phosphorylation and photoreduction in the guard cells requires much more investigation in a wider range of species and conditions. The possibility of other means of control by the mesophyll should not be ignored. In stressed conditions, however, the responses of g_s to vapour pressure deficit and leaf water potential ensure that g_s controls photosynthesis.

Hypothesis

The observations, experiments and arguments presented here are consistent with the following concept of stomatal action shown in Fig. 15.

1 In the short term, action is effected through processes occurring in the guard cells and associated epidermal cells. Translocation of metabolites from the mesophyll is only important for the longer term maintenance of resources.

2 Opening is the result of an increase in turgor relative to that of the surrounding cells. In some circumstances passive opening may occur as the result of loss of turgor in the surrounding cells. Active opening results from the accumulation of potassium ions in the vacuole against a concentration gradient. Potassium ion accumu-lation is the result of a proton pump(s) associated with the cell membrane(s). The charge balance in the vacuole is maintained by influx of chloride ion and the production of malate which also contribute to the osmotic potential. Water enters as a result of the reduction in total potential consequent upon the reduction in osmotic potential in the vacuole. Compatible solutes must also

accumulate in the cytoplasm to maintain osmotic equality with the vacuole.

3 Quanta are absorbed by chlorophyll in the guard cell chloroplasts where high-energy phosphate bonds and reduced adenine nicotinamide dinucleotide phosphate (NADPH$_2$) are produced as a result of photosynthetic electron transport. These or derived products are transported to the cell membranes where they are used to drive a proton/potassium ion pump.

4 Quanta are also absorbed by a blue-light absorbing pigment, possibly the flavin cryptochrome, which may be closely attached to the cell membranes. Quantum-driven electron transport either drives the proton/potassium ion pump directly or provides energy-rich and reduced products which are transported to the sites of action.

5 Tricarboxylic acid cycle respiration may also provide high-energy phosphate bonds and reducing power to drive the ion pump, in the dark for example.

6 The enzyme ribulosebisphosphate carboxylase/oxygenase

Fig. 15. A diagram of an energy requiring pump-leak system possibly involved in stomatal action. The primary interaction is through the availability of energy and reducing power to drive the potassium ion pump. This is provided by both chlorophyll and cryptochrome as well as by tricarboxylic acid cycle respiration (TCA). Other metabolic processes, including carboxylation by PEP carboxylase, compete for these products. Carbon dioxide is also shown as influencing leakage of potassium ions from the vacuole to the apoplasm.

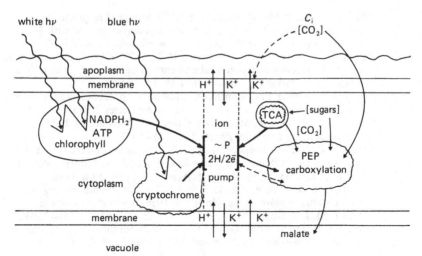

(Willmer, this volume) is currently thought to be absent from guard cells so that there is no photosynthetic carbon reduction cycle and no photorespiratory carbon oxidation cycle. Phosphoenol pyruvate (PEP) carboxylase fixes carbon dioxide into malate in the cytoplasm utilising some products of electron transport and both internally generated and externally supplied carbon dioxide.

7 It is only possible to speculate about the sites of action of carbon dioxide. Carbon dioxide may affect the ion pump and hence turgor by competing for high-energy phosphate bonds and reducing power for the PEP carboxylation. Thus high carbon dioxide and a high rate of carboxylation could lead to a reduced potassium ion influx. Alternatively, high carbon dioxide may increase the outward leakage of carbon dioxide from the vacuole to the cell exterior down the concentration gradient. Thus the net influx of potassium ion would be a function of carbon dioxide concentration at the membrane.

8 The system is highly regulated by specific ion concentrations, pH differences and effects of turgor on the activity of the ion pumps in ways which are only sketchily understood at present.

Mr J. I. L. Morison thanks the Natural Environmental Research Council for a studentship held during the preparation of this review.

References

Allaway, W. G. & Mansfield, T. A. (1967). Stomatal responses to changes in carbon dioxide concentration in leaves treated with 3-(4-chlorophenyl)-1, 1-dimethylurea. *New Phytologist* **66**, 57–63.

Beadle, C. L., Jarvis, P. G. & Neilson, R. E. (1979). Leaf conductance as related to xylem water potential and carbon dioxide concentration in Sitka spruce. *Physiologia Plantarum* **45**, 158–66.

Boyer, J. S. (1976). Photosynthesis at low water potentials. *Philosophical Transactions of the Royal Society of London* Series B **273**, 501–12.

Brunner, U. & Eller, B. M. (1974). Öffnen der Stomata bei hoher Temperatur im Dunkeln. *Planta* **121**, 293–302.

Burrows, F. J. & Milthorpe, F. L. (1976). Stomatal conductance in the control of gas exchange. In *Water Deficits and Plant Growth*, Vol. IV, ed. T. T. Kozlowski, pp. 103–52. Academic Press, New York.

Cowan, I. R. & Milthorpe, F. L. (1968). Plant factors influencing the water status of plant tissues. In *Water Deficits and Plant Growth*, Vol. I, ed. T. T. Kozlowski, pp. 137–193. Academic Press, New York.

Cowan, I. R. & Farquhar, G. D. (1977). Stomatal function in relation to leaf metabolism and environment. In *Integration of Activity in the*

Higher Plant. Society for Experimental Biology Symposium **31**, 471–505.

Dittrich, P. & Raschke, K. (1977). Malate metabolism in isolated epidermis of *Commelina communis* L. in relation to stomatal functioning. *Planta* **134**, 77–81.

Dubbe, D. R., Farquhar, G. D. & Raschke, K. (1978). Effect of abscisic acid on the gain of the feedback loop involving carbon dioxide and stomata. *Plant Physiology* **62**, 413–17.

Farquhar, G. D. (1978). Feedforward responses of stomata to humidity. *Australian Journal of Plant Physiology* **5**, 787–800.

Farquhar, G. D., Dubbe, D. R. & Raschke, K. (1978). Gain of the feedback loop involving carbon dioxide and stomata: theory and measurement. *Plant Physiology* **62**, 406–12.

Fischer, R. A. (1968). Stomatal opening in isolated epidermal strips of *Vicia faba* 1: response to light and to CO_2-free air. *Plant Physiology* **43**, 1947–52.

Gaastra, P. (1959). Photosynthesis of crop plants as influenced by light, carbon dioxide, temperature, and stomatal diffusion resistance. *Mededelingen van de Landbouwhogeschool te Wageningen, Nederland*, **59**, 1–68.

Goudriaan, J. & van Laar, H. H. (1978). Relations between leaf resistance, CO_2-concentration and CO_2-assimilation in maize, beans, lalang grass and sunflower. *Photosynthetica* **12**, 241–9.

Gradmann, H. (1928). Untersuchungen über die Wasserverhältnisse das Bodens als Grundlage des Pflanzenwachstums. *Jahrbuch für wissenschaftliche Botanik* **89**, 1–100.

Heath, O. V. S. (1948). Studies in stomatal action. Control of stomatal movement by a reduction in the normal carbon dioxide content of the air. *Nature* **161**, 179–81.

Heath, O. V. S. & Orchard (1957). Temperature effects on the minimum intercellular space carbon dioxide concentration "Γ". *Nature* **180**, 180–181.

Heath, O. V. S. & Russell, J. (1954). Studies in stomatal behaviour. VI. An investigation of the light responses of wheat stomata with the attempted elimination of control by the mesophyll. Part I. Effects of light independent of carbon dioxide and their transmission from one part of the leaf to another. *Journal of Experimental Botany* **5**, 1–15.

Honert, van den (1948). Water transport in plants as a catenary process. *Discussions of the Faraday Society* **3**, 146–53.

Hsiao, T. C. (1973). Plant responses to water stress. *Annual Review of Plant Physiology* **24**, 519–70.

Hsiao, T., Allaway, W. G. & Evans, L. T. (1973). Action spectra for guard cell Rb^+ uptake and stomatal opening in *Vicia faba*, *Plant Physiology* **51**, 82–8.

Hsiao, T. C. (1976). Stomatal ion transport. In *Transport in Plants*, *Encyclopedia of Plant Physiology New Series* **2 B**, pp. 195–221, eds U. Lüttge & N. G. Pitman. Springer-Verlag, Berlin, Heidelberg and New York.

Jarvis, P. G. (1971). The estimation of resistances to carbon dioxide transfer. In *Plant Photosynthetic Production/Manual of Methods*, eds Z. Šesták, J. Čatský, & P. G. Jarvis, pp. 566–631. Dr. W. Junk, N. V., The Hague.

Jarvis, P. G. (1975). Water transfer in plants. In *Heat and Mass Transfer in the Biosphere 1. Transfer Processes in Plant Environment,* ed. D. A. de Vries & N. H. Afgan, pp. 369–94. Scripta Book Co., Washington D.C.

Jarvis, P. G. (1980). Stomatal response to water stress in conifers. In *Adaptation of Plants to Water and High Temperature Stress,* ed. N. C. Turner & P. J. Kramer, pp. 105–122. John Wiley & Sons, New York, Chichester, Brisbane and Toronto.

Jarvis, P. G. (1981). Stomatal conductance, gaseous exchange and transpiration. In *Plants and their Atmospheric Environment,* ed. J. Grace, E. D. Ford & P. G. Jarvis, pp. 175–203. Blackwell Scientific Publications, Oxford.

Keck, R. W. & Boyer, J. S. (1974). Chloroplast response to low leaf water potentials. III. Differing inhibition of electron transport and photophosphorylation. *Plant Physiology* **53**, 474–9.

Kuiper, P. J. C. (1964). Dependence upon wavelength of stomatal movement in epidermal tissue of *Senecio odoris. Plant Physiology* **39**, 952–5.

Lake, J. V. (1967). Respiration of leaves during photosynthesis I. Estimates from an electrical analogue. *Australian Journal of Biological Science* **20**, 487–493.

Lake, J. V. & Slatyer, R. O. (1970). Respiration of leaves during photosynthesis. III. Respiration rate and mesophyll resistance in turgid cotton leaves, with stomatal control eliminated. *Australian Journal of Biological Science* **23**, 529–35.

Landsberg, J. J. & Butler, D. R. (1980). Stomatal response to humidity: implications for transpiration. *Plant, Cell and Environment* **3**, 29–33.

Leverenz, J. W. & Jarvis, P. G. (1979). Photosynthesis in Sitka spruce. VIII. The effects of light flux density and direction on the rate of net photosynthesis and the stomatal conductance of needles. *Journal of Applied Ecology* **16**, 919–32.

Leverenz, J. (1980). Photosynthesis and transpiration in large forest-grown Douglas-fir: diurnal variation. *Canadian Journal of Botany, in press.*

Lösch, R. (1977). Responses of stomata to environmental factors – experiments with isolated epidermal strips of *Polypodium vulgare.* I. Temperature and humidity. *Oecologia (Berl.)* **29**, 85–97.

Lösch, R. (1979). Responses of stomata to environmental factors – experiments with isolated epidermal strips of *Polypodium vulgare.* II. Leaf bulk water potential, air humidity, and temperature. *Oecologia (Berl.)* **39**, 229–38.

Mansfield, T. A. (1965). Studies in stomatal behaviour XII. Opening in high temperature in darkness. *Journal of Experimental Botany* **16**, 721–31.

Mansfield, T. A. & Meidner, H. (1966). Stomatal opening in light of different wavelengths: effects of blue light independent of carbon dioxide concentration. *Journal of Experimental Botany* **17**, 510–21.

Mansfield, T. A. (1976). Delay in the response of stomata to abscisic acid in CO_2-free air. *Journal of Experimental Botany* **27**, 559–64.

Meidner, H. & Heath, O. V. S. (1959). Stomatal responses to temperature and carbon dioxide concentration in *Allium cepa* L. and their relevance to midday closure. *Journal of Experimental Botany* **10**, 206–19.

Meidner, H. & Mansfield, T. A. (1968). *Physiology of Stomata.* 179 pp. McGraw-Hill, London.

Milthorpe, F. L. (1961). Plant factors involved in transpiration. In *Plant Water Relationships in Arid and Semi-arid Conditions. Arid Zone Research* 16, 107–15. UNESCO, Paris.

Monteith, J. L. (1965). Evaporation and environment. In *The State and Movement of Water in Living Organisms. Society for Experimental Biology Symposium* 19, 205–34.

Morison, J. I. L. (1980). Light and CO_2 effects on stomata. *Ph.D. Thesis, University of Edinburgh.*

Morris, J. G. (1968). *A Biologist's Physical Chemistry.* 367 pp. Arnold, London.

Mouravieff, I. (1965). Sur les réactions des çellules stomatiques au rayonnement ultraviolet proche en presence ou en absence du gaz carbonique. *Compte rendu hebdomadaire des séances de l'Academie des Sciences, Paris* 260, 5392–4.

Neilson, R. E. & Jarvis, P. G. (1975). Photosynthesis in Sitka spruce *(Picea sitchensis* (Bong.) Carr.) VI. Response of stomata to temperature. *Journal of Applied Ecology* 12, 879–89.

Ng, P. A. P. (1978). Response of stomata to environmental variables in *Pinus sylvestris* L. *Ph.D. Thesis, University of Edinburgh.*

Ng, P. A. P. & Jarvis, P. G. (1980). Hysteresis in the response of stomatal conductance in *Pinus sylvestris* L. needles to light: observations and a hypothesis. *Plant, Cell and Environment* 3, 207–16.

Pallas, J. E. (1965). Transpiration and stomatal opening with changes in carbon dioxide content of the air. *Science* 147, 171–3.

Parkinson, K. J. (1968). Apparatus for the simultaneous measurement of water vapour and carbon dioxide exchanges of single leaves. *Journal of Experimental Botany* 19, 840–56.

Penman, H. L. (1948). Natural evaporation from open water, bare soil and grass. *Proceedings of the Royal Society London* Series A 193, 120–45.

Philip, J. R. (1957). The physical principles of soil water movement during the irrigation cycle. *Proceedings of the III International Congress of Irrigation and Drainage* 8, 125–54.

Raschke, K. (1975). Simultaneous requirement of carbon dioxide and abscisic acid for stomatal closing in *Xanthium strumarium* L. *Planta* 125, 243–59.

Raschke, K. (1979). Movements using turgor mechanisms. In *Physiology of Movements, Encyclopedia of Plant Physiology New Series* 7, 383–441, eds W. Haupt & M. E. Feinleib. Springer-Verlag, Berlin, Heidelberg and New York.

Reed, K. L., Hamerly, E. R., Dinger, B. E. & Jarvis, P. G. (1976). An analytical model for field measurements of photosynthesis. *Journal of Applied Ecology* 13, 925–42.

Rogers, C. A., Powell, R. D. & Sharpe, J. H. (1979). Relationship of temperature to stomatal aperture and potassium accumulation in guard cells of *Vicia faba. Plant Physiology* 63, 388–91.

Rogers, C., Sharpe, P. J. H. & Powell, R. D. (1980). Dark opening of stomates of *Vicia faba* in CO_2-free air. *Plant Physiology* 65, 1036–38.

Schulze, E.-D., Lange, O. L., Buschbom, U., Kappen, L. & Evanari, M. (1972). Stomatal responses to changes in humidity in plants growing in the desert. *Planta* 108, 259–70.

278   P. G. JARVIS & J. I. L. MORISON

Schulze, E.-D., Lange, O. L., Evenari, M., Kappen, L. & Buschbom, U. (1974). The role of air humidity and leaf temperature in controlling stomatal resistance of *Prunus armeniaca* L. under desert conditions. I. A simulation of the daily course of stomatal resistance. *Oecologia (Berl.)* **17**, 159–70.

Schulze, E.-D., Lange, O. L., Kappen, L., Evenari, M. & Buschbom, U. (1975). The role of air humidity and leaf temperature in controlling stomatal resistance of *Prunus armeniaca* L. under desert conditions. II. The significance of leaf water status and internal carbon dioxide concentrations. *Oecologia (Berl.)* **18**, 219–33.

Sheriff, D. W. (1977). The effect of humidity on water uptake by and viscous flow resistance of excised leaves of a number of species: physiological and anatomical observations. *Journal of Experimental Botany* **28**, 1399–1407.

Sheriff, D. W. & Kaye, P. E. (1977). The response of diffusive conductance in wilted and unwilted *Atriplex hastata* L. leaves to humidity. *Zeitschrift für Pflanzenphysiologie* **83**, 463–6.

Thornley, J. M. M. (1976). *Mathematical Models in Plant Physiology.* pp. 318. Academic Press, London, New York & San Francisco.

Travis, A. J. & Mansfield, T. A. (1977). Studies of malate formation in 'isolated' guard cells. *New Phytologist* **78**, 541–6.

Travis, A. J. & Mansfield, T. A. (1979). Stomatal responses to light and CO_2 are dependent on KCl concentration. *Plant, Cell and Environment* **2**, 319–23.

Turner, N. C. (1979). Drought resistance and adaptation to water deficits in crop plants. In *Stress Physiology in Crop Plants,* ed. H. Mussell & R. C. Staples, pp. 344–372. John Wiley & Sons, New York, Chichester, Brisbane and Toronto.

Tyree, M. T. & Yianoulis. P. (1980). The site of water evaporation from substomatal cavities, liquid path resistances and hydroactive stomatal closure. *Annals of Botany,* **46**, 175–93.

Unsworth, M. H., Biscoe, P. V. & Black, V. (1976). Analysis of gas exchange between plants and polluted atmospheres. In *Effects of Air Pollutants on Plants, S. E. B. Seminar Series 1,* ed. T. A. Mansfield, pp. 5–16. Cambridge University Press, London, New York & Melbourne.

Warrit, B. (1977). Studies on stomatal behaviour in apple leaves. *Ph.D. Thesis, University of Bristol.*

Warrit, B., Landsberg, J. J. & Thorpe, M. R. (1980). Responses of apple leaf stomata to environmental factors. *Plant, Cell and Environment* **3**, 13–22.

Watts, W. R., & Neilson, R. E. (1978). Photosynthesis in Sitka spruce (*Picea sitchensis* (Bong.) Carr.) VIII. Measurements of stomatal conductance and $^{14}CO_2$ uptake in controlled environments. *Journal of Applied Ecology* **15**, 245–55.

Weatherley, P. E. (1970). Some aspects of water relations. *Advances in Botanical Research* **3**, 171–206.

Willmer, C. M., Thorpe, N., Rutter, J. C. & Milthorpe, F. L. (1978). Stomatal metabolism: carbon dioxide fixation in attached and detached epidermis of *Commelina. Australian Journal of Plant Physiology* **5**, 767–78.

Wong, S. C., Cowan, I. R. & Farquhar, G. D. (1978). Leaf conductance in relation to assimilation in *Eucalyptus pauciflora* Sieb. ex Spreng.

Influence of irradiance and partial pressure of carbon dioxide. *Plant Physiology* **62**, 670–4.

Wong, S. C., Cowan, I. R. & Farquhar, G. D. (1979). Stomatal conductance correlates with photosynthetic capacity. *Nature* **282**, 424–76.

Zeiger, E. & Hepler, P. K. (1977). Light and stomatal function: blue light stimulates swelling of guard cell protoplasts. *Science* **196**, 887–8.

H. MEIDNER

What next?

In commenting on some of the topics that were discussed during the seminar I shall have to move from one contribution to another. I may be forgiven therefore if the subject matter changes abruptly and there are some unanswered questions.

Development and anatomy

Stomatal aperture is determined by the shape of the guard cells. This was appropriately referred to in the opening discussion about development, morphology and anatomy which was necessary for an understanding of the functioning of guard cells. We were given details about the shaping of the differentiating guard cells which revealed that in addition to genetically controlled processes within guard cells, subsidiary cell turgor and environmental conditions contribute to the fine structure and shaping of the guard cell walls. Thus the observation that the formation of subsidiary cells always precedes guard cell formation assumes added significance as do the properties of cell walls which allow for the establishment of plant cell turgor pressure. In this context we were shown polarizing microscope pictures of guard cells of *Cyperus esculentus* L. with their middle portion micellated either longitudinally or radiating out from the pore. This is ideal material for testing the validity of the generally accepted conclusions based on Ziegenspeck's fundamental observations of micellation (1955), and I mean to carry out this test soon. The optical microscope is still the physiologist's best friend.

There is further work to be done. The apoplast, besides being involved in developmental cell shaping, is clearly important in connection with other aspects of stomatal functioning. These topics are dear to my heart and I can mention some specific matters that arose during the discussions and require elucidation.

    (1) The chemical composition and properties of the walls of the cells of the stomatal apparatus and their different cuticular coverings.

(2) The anatomical and functional connections between mesophyll tissue and the epidermis which make possible, among other things, the synthesis, often by chloroplast-free cells, of a great variety of complex compounds that serve to build up the external and internal cuticular layers. To what extent is the translocation of metabolites between mesophyll and epidermis functionally important to stomata?

(3) The physiological role of changes in the turgor pressure of guard cells as influencing membrane potentials and translocation processes into and out of guard cells and into the cuticle.

(4) The implications for guard cell functioning of the changes in pH of the water in the guard cell walls, i.e. the medium bathing the guard cells *in situ*. These changes depend on proton extrusion, solubility of carbon dioxide (and pollutants?) and possibly on changes in the matric potential of walls.

(5) Changes in the concentrations of other solutes that move or are deposited in the cell walls. I think the microchamber which was described to us and the technique of micro-surgically excising guard cells open the way for exploring this matter.

Guard cell metabolism

Some of the topics referred to so far lead us on to the metabolism of guard cells. The question was asked: Is it perhaps correct to identify a primary process in metabolic guard cell action and is proton extrusion such a primary process? This question will have to be answered and others such as the effects of low or very low pH (5.5–3.0), the action of fusicoccin, and the movements of ions with and without their counter ions across plasmalemma and tonoplast. Possible changes in the properties of cell walls due to changes in pH are another topic for experimentation.

Quantitative studies are needed of the changes that occur in the cytoplasm, especially those affecting carbohydrate metabolism, including that of starch, which is apparently involved in the much discussed humidity response. Also, enzyme activities leading to ATP and NAD formation which were the subject of earlier but incomplete investigations (Fujino, 1967) should be looked at once more. Changes associated with the morphology of cytoplasmic organelles (Heller & Resch, 1967; Guyot & Humbert, 1970; Heller, Kausch & Trapp, 1971; Humbert & Guyot, 1972; Guyot, Humbert & Louguet, 1975) and changes in the water relations between cytoplasm and vacuole need to be taken into account also (Sheriff & Meidner, 1975). Since metabolic activity in the cytoplasm involves events at the plasmalemma it is important to realize that such processes would cease unless they were linked with others occurring at the tonoplast. Do isolated vacuoles provide improved material for such studies?

Guard cell metabolism as affected by abscisic acid (ABA) continues to pose questions. ABA is present in quantity in mesophyll chloroplasts. Stomata are known to respond to traces of ABA, and exogenously supplied ABA accumulates preferentially in the cells of the stomatal complex. Does ABA initiate stomatal closure in response to stress or does it reinforce closure initiated hydropassively? Is the internal redistribution of ABA more important than the total amount present, and can we detect those traces of ABA that may initiate stomatal closure? Is the redistribution of ABA linked to its postulated involvement in phloem loading (Malek & Baker, 1978) when leaf water stress develops beyond that existing during normal transpiration? Is the postulated redistribution of ABA a distinct process or is it the result of changes in epidermal cell water content, i.e. is there ABA 'redistribution' or 'increased concentration' due to changes in the epidermal water content? In most leaves such changes are more pronounced in the epidermis than in the mesophyll tissue (Meidner, 1952).

Stomatal responses to stimuli

The involvement of ABA in responses to specific stimuli leads me now to mention other aspects of stomatal movements. Allied to studies of responses is a topic which requires further investigation, namely, the kinetics of guard cell deformations. The rate, magnitude and direction of the movements in response to stimuli must be measured. There are direct links between the dynamics of movement and some responses to specific environmental variables. As long ago as 1950 Williams and Heath found that high atmospheric humidity slowed down responses to light and carbon dioxide, thus hinting at a link with studies of epidermal water relations which have by 1980 developed to include the concepts of capacitance and cell wall matric potential.

A different kind of dynamic study has recently been reported. Detached leaves of *Avena sativa* L. were exposed to regularly applied light/dark perturbations producing alternating pulse responses in transpiration rate (Johnsson & Skaar, 1979). These responses were found to be reinforced by the presence of sodium chloride in the transpiration stream. Has this observation any significance in connection with the reported lessening of light, carbon dioxide and ABA effects by the substitution of sodium for potassium in an epidermal incubation medium?

Contradictory reports on the nature of stomatal responses to stimuli have existed for some time and it is well to remember that this could be due to different experimental methods. These are usually reported nowadays in small print but they give essential information if valid conclusions are to be drawn from experiments. These 'Materials and Methods' sections must include the definition of the type of plant material used, the point reached in the endogenous rhythm of stomatal behaviour and, as we have heard again, such matters as height of insertion of the leaf – in short its previous history. All of these assume significance if we are to consider that some stimuli may be triggers setting free motors which are prepared to respond and are not themselves suppliers of the energy which sustains stomatal movements.

Stomatal control of processes

Last, we come to the question of whether stomata exert control over transpirational vapour loss and photosynthetic carbon dioxide intake or whether these processes largely determine the degree of stomatal opening. Feedback mechanisms have been postulated and for many years a degree of mutual interdependence has been recognized between stomatal functioning and the gas exchange processes which are largely regulated by stomatal conductance. However, a distinction should be emphasized between primary functional mechanisms such as metabolism or turgor changes which bring about guard cell deformation and secondary control functions that can be postulated. The latter are often based on theoretical models for which adequate data do not as yet exist. Their logic is sometimes circular as can be illustrated by reference to the oft quoted equation defining the internal concentration of carbon dioxide. This equation employs a value for leaf conductance which is calculated from transpiration measurements, i.e. a value which even for transpirational vapour loss is questionable because it depends on the almost certainly erroneous assumption that the leaf airspace at the evaporation sites is practically saturated with water vapour. This calculated value for leaf conductance in respect of water vapour is then used for the calculation of the internal carbon dioxide concentration in which carbon dioxide transfer is treated in the same way as water vapour transfer although their diffusion paths are not the same. The situation arises then that a model is used to support a postulate, e.g. a mechanism for the humidity response, and the hypothesis serves to support the model, whereas what is needed is the experimental test of both hypothesis and model. Thus, a postulated 'feed forward' mechanism in respect of the humidity response lacks evidence of a truly peristomatal controlling sensing mechanism outside the stomatal throat which would distinguish it from a response to water stress in the cells surrounding the sub-stomatal cavity. Could the reported unlignified evaporative sites in the pine needle subsidiary cell (Esau, 1977) serve as such sensors and are they not cutinized?

Good modelling is of value for the elucidation of control mechanisms and it allows us to identify the gaps in our information which require filling in. Thus the role of models is in the researching of secondary control functions that govern guard cell metabolism.

To conclude, let me reaffirm my faith in experimental stomatal physiology by returning to 'Materials and Methods'. Doubts have been expressed about the validity of results obtained, for instance, with epidermal strips for the interpretation of events in intact leaves. Such doubts are justified but they should not lead us to dismiss those results. There is a place in experimental work in stomatal physiology for all materials including leaf canopies (unless information is desired that is applicable directly to stomatal movements), intact leaves, epidermal strips, and isolated parts such as stomata, guard cells, protoplasts and vacuoles. Likewise all experimental methods have their application. However, two privisos should be noted. To arrive at generally valid conclusions, observations made in different materials must be quantitatively comparable as to rate and magnitude of response, and, if it is proposed to offer a common interpretation of results obtained by different measuring techniques, this must be justified.

References

Esau, K. (1977). *Anatomy of Seed Plants.* New York: John Wiley and Sons.

Fujino, M. (1967). Role of ATP and ATPase in stomatal movement. *Science Bulletin of the Faculty of Education, Nagasaki University* **18**, 1–47.

Guyot, M. & Humbert, C., (1970). Les modifications du vacuome des cellules stomatique. *Comptes Rendus de l'Academie des Sciences, Paris* **270**, 2787–90.

Guyot, M., Humbert, C. & Kouguet, P. (1975). Etude ultrastructurales comparée des cellules stomatique d'ouverture ou de fermeture. *Comptes Rendus de l'Academie des Sciences, Paris* **280**, 1373–6.

Heath, O. V. S. (1950). The role of carbon dioxide in the light response of stomata. *Journal of Experimental Botany* **1**, 29–61.

Heller, F. O. & Resch, A. (1967). Funktionell bedingter Strukturwechsel der Zellkerne in den Schliesszellen von *Vicia faba*. *Planta* **75**, 243–52.

Heller, F. O., Kausch, W. & Trapp, L. (1971). U-V mikroskopischer Nachweis von Strukturveränderungen in Schliesszellen von *Vicia faba*. *Naturwissenschaften* **58**, 419.

Humbert, C. & Guyot, M. (1972). Modifications structurelles des cellules stomatique. *Comptes Rendus de l'Academie des Sciences, Paris* **274**, 380–2.

Johnsson A. & Skaar, H. (1979). Alternating perturbation responses of the regulatory system in *Avena* leaves. *Physiologia Plantarum* **46**, 218–20.

Malek, F. & Baker, D. A. (1978). Effect of fusicoccin on proton
   co-transport of sugars in phloem loading. *Plant Science Letters* **11**,
   233–9.
Meidner, H. (1952). An instrument for the continuous determination of
   leaf thickness changes. *Journal of Experimental Botany* **3**, 319–25.
Sheriff, D. W. & Meidner, H. (1975). Correlation between the unbound
   water content of guard cells and stomatal aperture. *Journal of
   Experimental Botany* **26**, 315–18.
Williams, W. T. (1950). The water relations of the epidermis. *Journal of
   Experimental Botany* **1**, 114–31.
Ziegenspeck, H. (1955). Das Vorkommen von Fila in radialer
   Anordnung in den Schliesszellen. *Protoplasma* **44**, 385–92.

INDEX

abscisic acid (ABA), 29, 63, 77, 98,
129, 209, 265, 283
[^{14}C], 170
diurnal variation, 171
effect on wilted *Pisum* leaves, 171
field-grown plants, 172
and guard cell ion movements, 66, 67
production and leaf turgor, 172, 173
responses and CO_2, 130
role in drought, 174–8
and root and shoot growth, 177
and root water uptake, 175
site of synthesis, 168
and stomatal behaviour, 168–73
stomatal closure, 38, 97, 107, 167, 170
and stomatal dependence, 38
and water stress, 137, 167–73, 177, 178
as water stress substitute, 173, 174
Acer saccharum, 141
Aeonium glutinosum
stomatal opening anomaly, 139
Aeonium spathulatum
stomata and humidity, 139
Aeonium urbicum
stomata and humidity, 139
age and stomatal response, 233
aerodynamic conductance, 226
Aichryson dumosum
hairyness, 155, 156
stomata and humidity, 139, 155
Aichryson laxum
hairyness, 156
stomata and humidity, 139, 155, 156
Aichryson villosum
hairyness, 155, 156
stomata and humidity, 139, 155
alanine aminotransferase, 89, 90
in guard cells, 94
Allium cepa, 22
anions in stomata, 76, 77
Calvin cycle in epidermis, 95
chloride in guard cells, 99
chloroplast fluorescence, 110
guard cell metabolism, 98

guard cell thickening, 3–5
guard mother cell division, 2
malate accumulation, 77, 99; role of, 99
reserve carbohydrates, 76, 98
vacuolar fluorescence, 113
wall deposition and colchicine, 16
Allium porrum, 99
Alternaria tenuis
toxin and stomatal behaviour, 211, 212
Amaranthus powelli, 263
amphistomatous leaves, 37
porometer clamps, 43
Anemia rotundifolia
cytoplasm appearance and stomatal
opening, 80
anions
guard cell cytoplasm and organelles, 80,
81
light and CO_2 effects on formation, 130–3
organic and potassium change, 53
in stomatal operation, 71–81
antitranspirants, *see* ABA and farnesol
Aphanomyces cochlioides
sugarbeet wilt, 208
Arachis hypogaea, 230
argon, 39
aspartate, 79
epidermal, 93
aspartate aminotransferase, 88–90
in guard cells, 94
ATP, guard cell production, 123
auramine O, cuticle fluorescence, 20, 21
Avena sativa, 263, 283
ozone response, 188, 189

benzo-18-crown-6
potassium binding, 129
Betula verrucosa
stomatal response to ABA and water
stress, 174, 175
bromine, [^{82}Br], 65, 66
buffer and guard cell studies, 29

Calluna vulgaris microclimate, 155